Principles and Practices for Diesel Contaminated Soils, Vol. III

Principles and Practices for Diesel Contaminated Soils, Vol. III

Edited by

PAUL T. KOSTECKI

EDWARD J. CALABRESE and

CHRISTOPHER P.L. BARKAN

Amherst Scientific Publishers

Copyright © 1994 by
Amherst Scientific Publishers

Reproduction of any part of this book without
the permission of Amherst Scientific Publishers
is unlawful. Requests for permission or
additional information should be addressed to
Amherst Scientific Publishers, P.O. Box 2308,
Amherst, MA 01002

Printed in the United States of America

ISBN 1-884940-01-3

Biographies

Edward J. Calabrese is a board certified toxicologist who is professor of toxicology at the University of Massachusetts School of Public Health, Amherst. Dr. Calabrese has researched extensively in the area of host factors affecting susceptibility to pollutants, and is the author of more than 270 papers in scholarly journals, as well as 23 books, including *Principles of Animal Extrapolation; Nutrition and Environmental Health*, Vols 1 and 2: *Ecogenetics: Safe Drinking Water Act: Amendments, Regulations and Standards; Petroleum Contaminated Soils*, Vols, 1, 2 and 3; *Ozone Risk Communications and Management; Hydrocarbon Contaminated Soils*, Vols. 1, 2, and 3; *Hydrocarbon Contaminated Soils and Groundwater*, Vols. 1, 2 and 3; *Multiple Chemical Interactions; Air Toxics and Risk Assessment; Alcohol Interactions with Drugs and Chemicals; Regulating Drinking Water Quality; Biological Effects of Low Level Exposures to Chemicals and Radiation; Contaminated Soils; Diesel Fuel Contamination; Risk Assessment and Environmental Fate Methodologies, and Principles and Practices for Petroleum Contaminated Soils*, Vols. 1 and 2. He has been a member of the U.S. National Academy of Sciences and NATO Countries Safe Drinking Water Committees, and the Board of Scientific Counselors for the Agency for Toxic Substances and Disease Registry (ATSDR). Dr. Calabrese also serves as Chairman of the International Society of Regulatory Toxicology and Pharmacology's Council for Health and Environmental Safety of Soils (CHESS) and Director of the Northeast Regional Environmental Public Health Center at the University of Massachusetts.

Paul T. Kostecki, Associate Director, Northeast Regional Environmental Public Health Center, School of Public Health, University of Massachusetts at Amherst, received his Ph.D. from the School of Natural Resources at the University of Michigan in 1980. He has been involved with risk assessment and risk management research for contaminated soils for the last eight years, and is coauthor of *Remedial Technologies for Leaking Underground Storage Tanks*, coeditor of *Soils Contaminated by Petroleum Products and Petroleum Contaminated Soils*, Vols. 1, 2 and 3, and of *Hydrocarbon Contaminated Soils*: Vols. 1, 2 and 3; and *Hydrocarbon Contaminated Soils and Groundwater*, Vols 1, 2 and 3, of *Contaminated Soils: Diesel Fuel Contamination; Principles and Practices for Petroleum Contaminated Soils*; and *Risk Assessment and Environmental Fate Methodologies*, Vols. 1 and 2. Dr. Kostecki's yearly conferences on hydrocarbon contaminated soils draw hundreds of researchers and regulatory scientists to present and discuss state-of-the art solutions to the multidisciplinary problems surrounding this issue. Dr. Kostecki also serves as Managing Director for the International Society of Regulatory Toxicology and Pharmacology's Council for Health and Environmental Safety of Soils (CHESS), as Executive Director of the Association for the Environmental Health of Soils (AEHS), and as Editorial Advisor to the *Journal of Soil Contamination* and *SOILS* magazine.

Christopher Barkan manages the Environmental and Hazardous Materials Research Program of the Association of American Railroads (AAR). The program sponsors and conducts research for the railroad industry on pollution prevention, remediation technologies and hazardous materials transportation safety. Dr. Barkan also serves as Deputy Project Director of the RPI-AAR Railroad Tank Car Safety Research and Test Project, a cooperative program of the tank car and railroad industries studying tank car safety. Barkan holds graduate degrees in biology from the State University of New York at Albany where he conducted research on the application of stochastic optimization models to ecological processes. He continued this work as a research fellow at the Smithsonian Institution's Environmental Research Center prior to joining the AAR Research and Test Department in 1988. His research at the AAR has included the development and application of quantitative risk assessment models for rail transportation of hazardous materials, with particular emphasis on the environmental impact of chemical spills. More recent research has included investigation of the environmental impact of railroad rail lubrication, a study of the pollutant levels in stormwater runoff from railroad facilities covered by the new EPA stormwater permit regulations, a review of industry results on toxicity characteristic leaching procedure testing of railroad crossties, and review and evaluation of the factors affecting successful application of *in situ* bioremediation for the railroad industry. His current research includes analyses of the effectiveness and the costs and benefits of options to reduce the environmental impact of various railroad operations and the integration of more risk-based approaches to remediation of railroad property contaminated with petroleum hydrocarbons. He is an author or editor on a number of papers, reports and books on these subjects

Preface

Proceedings of 1993 Symposium on Remediation of
Diesel Fuel Contaminated Soil

University of Massachusetts

*Sponsored by the Association for the American Railroads
Research and Test Department*

As the cost of remediation of contaminated soil on railroad property continues to grow, so does the importance of applying the most cost-effective approach to cleanup of sites. By their nature, railroads traverse large expanses of territory and in so doing they encounter a wide variety of soil and hydrological conditions that affect remediation. On the other hand, the contaminants of concern are fairly consistent. The most commonly encountered contaminant at railroad sites is diesel fuel, frequently in combination with other petroleum hydrocarbons such as lubricating oils or heavier fuels such as Bunker C. By contrast with many sites outside the railroad industry, gasoline and its constituents are relatively uncommon on railroad property. Consequently the railroad environmental officer is interested in how to remove a limited set of petroleum hydrocarbons, but under a wide variety of conditions. One of the objectives of this book is to provide the reader with the methods and results of approaches that have been used at a wide variety of sites, cleaning up contaminants of interest to railroads such as diesel and related compounds.

The general need to cleanup petroleum hydrocarbons such as weathered diesel and other heavy hydrocarbons leads to a related issue which is the risk posed by these materials. In general, the toxicity and/or the environmental mobility of compounds typical of railroad remediation sites is relatively low. Complicating the issue is that these contaminants are often encountered in conjunction with coal dust and cinders that have been deposited as a result of railroad operations over the past 150 years. The most commonly used analytical techniques, involving measurement of total petroleum hydrocarbons (TPH), are confounded by these other materials, as well as many other benign organic materials. As a result of reliance on TPH analyses, estimates of the risk posed by the contaminants of concern at many railroad remediation sites may be overestimated. Furthermore, many railroad remediation sites are located on property relatively distant from human habitation; therefore, the exposure of people to the contaminants is often low.

The timelines and disturbance resulting from site cleanup on railroad property are also issues. The urgency to clean up some sites may be relatively low. In these cases, lower cost, but possibly slower methods may be the most cost-effective approach. Under other circumstances, for instance in relation to a real estate transaction involving railroad property, time may be of the essence and a very aggressive approach to a rapid and punctual cleanup can be imperative. A complicating factor for many railroad remediation sites is that they may involve the cleanup of soil beneath active rail facilities. Under these circumstances a high value may be placed on *in situ* approaches in which treatment of the soil can take place without its removal.

Taken together, these factors point to the need for several considerations when approaching cleanups at railroad sites. One is the need to carefully match the cleanup technology with the local site conditions, including a thorough site assessment. Another is to take into account the background levels at a site before setting cleanup target levels. Finally, and perhaps most important,especially at large sites, is the need to evaluate the risk to human and environmental health due to the contaminants actually present at the site and the capability of the analytical techniques to allow measurement of that risk. The chapters included in this book address these issues and are based on presentations made at the symposium on remediation of diesel fuel contaminated soil sponsored by the Association of American Railroads, and held in conjunction with the Eighth Annual Contaminated Soils Conference organized by the University of Massachusetts in September 1993. It is hoped that the material presented herein will be of value to both the regulator and the regulated community in accomplishing more timely and cost-effective cleanup of sites contaminated with diesel fuel and other petroleum hydrocarbons.

Christopher P.L. Barkan
Manager
Environmental and Hazardous Materials Research Program
Research and Test Department
Association of American Railroads

Acknowledgements

This book could not have been produced without the continued interest and support of the railroad environmental community, in particular the members of the AAR Environmental Affair Committee and its Environmental Engineering and Operations Subcommittee. Their continued participation in the conference ensures the vitality and relevance of the material presented. Special thanks to Martina Schlauch for her assistance with the technical preparation of the conference and book and to Linda Rosen for her energetic and unceasing efforts to ensure the successful outcome of the conference.

Contents

CHAPTER 1

Health-Based Cleanup Goal for Diesel-Contaminated Soil - A Case Study of the No-Action Remedial Alternative for an Underground Storage Tank Site

Xuannga P. Mahini

Ogden Environmental and Energy Services, Inc. — San Francisco, California

INTRODUCTION

In February 1991, during the construction of a new building complex at a military facility in Hawaii, a backhoe damaged a previously unknown 1,000-gallon underground storage tank (UST) containing a dark liquid that consisted of water and an 8-inch layer of weathered diesel fuel.[1] The liquid and tank were removed and disposed of together with an adjacent empty 350-gallon UST found during the removal. Soil samples taken beneath the USTs detected total petroleum hydrocarbons as diesel (TPH-diesel) at concentrations of 5,690 parts per million (ppm) and low levels of TPH-diesel-related volatile organic compounds (VOC), such as ethylbenzene (140 parts per billion [ppb]), toluene (54 ppb), and xylenes (100 ppb). The maximum concentrations of these VOCs were below the 1991 Hawaii Department of Health's (HDOH) recommended soil cleanup goals and were, therefore, not of concern. TPH-diesel concentrations in the soil, however, exceeded the 1991 HDOH's recommended cleanup goal of 50 ppm.[2] Initial cleanup efforts conducted in May 1991 consisted of removal of the upper layers of the diesel-contaminated soil, resulting in an excavation of approximately 8 by 15 feet in area and 5 feet in depth.[3]

In August 1991, a site characterization study of the subsurface TPH-diesel release was performed at the site[4]. At that time, the 1991 HDOH cleanup goals for TPH-diesel, while recommended, were not enforceable (in 1992, HDOH revised the recommended cleanup criteria list, deleting TPHs as a criterion while adding selected semivolatile organic compounds [SVOC]). Based on an evaluation of site-specific conditions, such as contaminant distribution, geology, and land uses, a preliminary risk assessment (PRA), although not included in the original scope of work, was conducted to determine if diesel-contaminated soil, in excess of the 1991 HDOH recommended cleanup goal, would require an immediate removal or remediation prior to the completion of construction activities, and to establish alternate action levels that are protective of human health and the environment.[2] The PRA was conducted at no extra cost to the government and under an expedited schedule to minimize delays to the on-going construction project. It proved instrumental in supporting a no-action remedial alternative for diesel-contaminated soils at the site.

SITE DESCRIPTION

During the site characterization process, 19 soil samples from six borings (SB-1 to SB-6) at depths up to 40 feet below ground surface (bgs) and six grab groundwater samples were collected and submitted to an on-site mobile laboratory for TPH analysis. Figure 1-1 depicts the locations of the soil borings drilled at the site. Table 1-1 shows the results of TPH analysis, with a maximum TPH-diesel concentration of 800 ppm in the soils. Identified diesel-contaminated soil and grab groundwater samples were then sent to an off-site analytical laboratory for volatile and semivolatile organic analyses.[4] Analytical results showed that VOCs commonly associated with petroleum products, such as benzene, toluene, ethylbenzene, and xylenes (BTEX), were not detected in site soil samples.[4] The only VOCs detected in grab groundwater were xylenes at a trace level of 9 ppb. SVOCs, such as noncarcinogenic polynuclear aromatic hydrocarbons (nPAH), were detected only in the SB-1 soil at 2.0 feet below the bottom of the tank excavation, or at a depth of 7 feet bgs. All detected nPAH concentrations were below sample quantitation limits and were qualified as estimated (J) (none of these nPAH detections exceed the 1992 HDOH recommended cleanup criteria). The detected nPAHs at SB-1 were acenaphthene ($C_{12}H_{10}$ — 1,100 ppb), dibenzofuran ($C_{12}H_8O$ — 1,100 ppb), fluoranthene ($C_{16}H_{10}$ — 2,700 ppb), phenanthrene ($C_{14}H_{10}$ — 5,500 ppb), and pyrene ($C_{16}H_{10}$ — 2,100 ppb).

A review of the site conditions and contaminant distribution indicated that TPH-diesel levels in the soil, while exceeding the 1991 HDOH cleanup goal, appeared to be limited to depths of approximately 21 to 28 feet bgs (below a 20-foot basalt layer), which is the approximate depth of the water table (Table 1-1). These results suggested that if the abandoned USTs were the source of the contaminants, the released diesel fuel may have moved vertically downward along joints and fractures in the upper basalt layer to the water table, then spread laterally along the interface of the basalt layer and the underlying silty clay layer. Approximately 1.5 inches of light nonaqueous phase liquid (LNAPL) was measured in one soil boring (SB-6). Tentatively identified compounds (TIC) detected at the site further confirm the presence of TPH-diesel in soil and groundwater. Most of these TICs are common components of diesel fuel and fall into two chemical classes: aromatic compounds with 2-4 carbons in alkyl groups attached to the benzene ring (for example, diethyl benzene) and saturated aliphatic hydrocarbons in the C7 to C17 range (n-alkanes such as dodecane and hexadecane, as well as their methylated derivatives)[4]. Therefore, for purposes of this PRA, TPH-diesel and nPAHs are considered the potential contaminants of concern (COC) at the site.

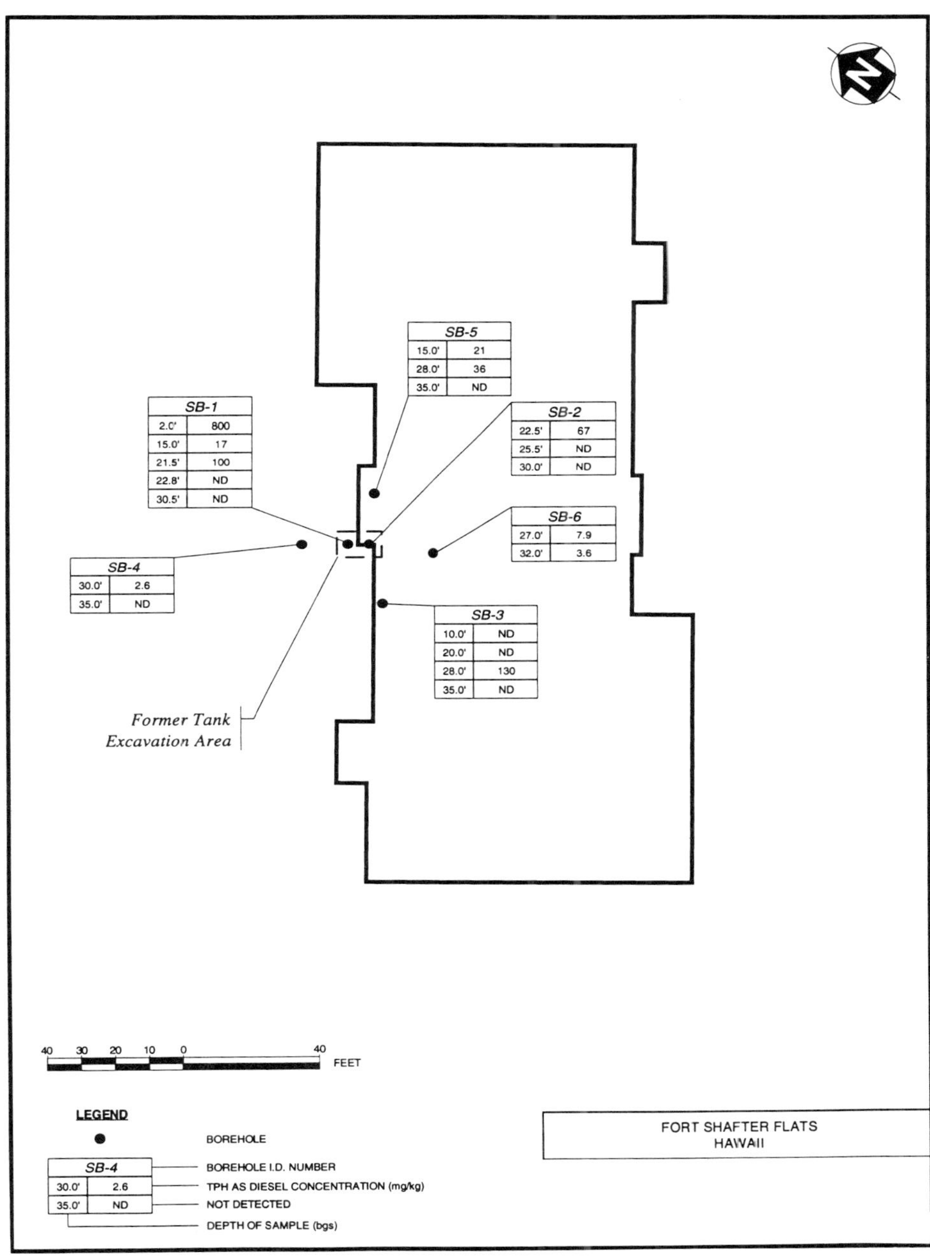

Figure 1-1   Concentrations of TPH-Diesel in Soil[4]

Table 1-1   Results of Total Petroleum Hydrocarbon Analysis

Sample No.	TPH[a]	Lithology
SOIL[b]		
SB-1-2.0	800	Silty Sand
SB-1-15.0	17	Basalt
SB-1-21.5	100	Gravelly Clay
SB-1-22.8	<2.5[c]	Silty Clay
SB-1-30.5	<2.5	Silty Clay
SB-2-22.5	6.7	Silty Clay
SB-2-25.5	<2.5	Silty Clay
SB-2-30.0	<2.5	Silty Clay
SB-3-10.0	<2.5	Basalt
SB-3-20.0	<2.5	Silty Clay
SB-3-28.0	130	Silty Clay
SB-3-35.0	<2.5	Silty Clay
SB-4-30.0	2.6	Silty Clay
SB-4-35.0	<2.5	Silty Clay
SB-5-15.0	21	Basalt
SB-5-28.0	36	Silty Clay
SB-5-35.0	<2.5	Silty Clay
SB-6-27.0	7.9	Silty Clay
SB-6-32.0	3.6	Silty Clay
Grab Groundwater		
GW-1	450	
GW-2	0.65	
GW-3	180	
GW-4	120	
GW-5	7.5	
GW-6	250	

Notes:

a TPH concentrations (as diesel) were analyzed in accordance with the California Leaking Underground Fuel Tank (LUFT) Field Manual Methodology.[104] Concentrations are in mg/kg (ppm) for soil samples and mg/L (ppm) for groundwater samples.

b Value at the end of a soil sample number denotes the depth, in feet, below ground surface (bgs) where sample was taken. For example, sample SB-1-22.8 was collected at 22.8 feet bgs.

c "<" indicates concentrations below sample quantitation limits.

RISK ASSESSMENT METHODOLOGY

The absence of VOCs and the presence of low levels of nPAHs, coupled with the lack of complete soil exposure pathways at the site, suggested that subsurface soil contamination at the site would pose little hazard to human health or the environment. For example, once the excavation is backfilled and the area is covered by a planned walkway, no potential for direct contact with the contaminated soil is anticipated. Furthermore, the use of the U.S. Environmental Protection Agency (EPA) standard risk assessment methodology, which evaluates only hazardous constituents of mixtures, such as nPAHs, would not fully reflect the potential risks posed by TPH-diesel at the site.

To confirm the above assumptions on a quantitative basis, toxicity values were derived for diesel fuel as a complex mixture. The derivations were based on currently available toxicological data on diesel fuel and similar mixtures and current scientific knowledge of human risk assessment methodology.[5-14] To be health conservative, the toxicity values of diesel fuel as a mixture were used in conjunction with toxicity values for known toxic constituents of diesel, such as nPAHs, to evaluate the point estimate reasonable maximum exposure (RME) and the potential human health risks posed by COCs via three common soil-related exposure routes: ingestion, inhalation of particulates, and dermal contact. Whenever possible, the EPA standard default factors for residential scenarios[15,16] were used in the calculation of risk values and health-based cleanup goals (HBG), because these factors are generally more health protective.

The basis for evaluating TPH-diesel as a mixture is outlined in the current National Research Council (NRC) and EPA guidelines,[7,9,13,14] which recommend that for the risk assessment of complex mixtures (such as diesel fuel) data on the mixture as an entity or on a "sufficiently similar" mixture may be used. The criteria for determining "sufficient similarity" are intentionally vague and are likely to vary on a case-by-case basis, depending on the nature and quality of the available data and the extent of human exposure.[9] In general, the determination of "sufficient similarity" takes into consideration the variation in chemical composition and toxicologic properties of the mixture of interest and similar mixtures. A review of the available literature revealed that middle distillate fuel products, such as kerosene, jet fuel oil, and hydrodesulfurized middle distillates, display similar types and degrees of toxicity as diesel fuel. A discussion of diesel fuel and similar middle distillate mixtures, therefore, will be presented throughout this paper.

HAZARD IDENTIFICATION

Hazard identification is the first step in the health risk assessment process. It includes a description of the risk agent's potential for causing adverse health or environmental effects. This section addresses the hazard caused by exposure to diesel fuel as a mixture and nPAHs.

Diesel Fuel as a Mixture

This section briefly presents all factors that determine or affect the degree of hazard caused by exposure to diesel fuel, including fuel composition, physical and chemical properties, environmental fate and transport characteristics, and available health and environmental effects information.[17]

Types and Composition of Diesel Fuel

Virgin diesel fuel or fuel oil is a clear, colorless to pale yellow liquid with a mild petroleum odor and a semivolatile nature. It is a complex hydrocarbon mixture derived from the middle distillate fraction of petroleum, which has a boiling range of 350 to 700 °F (or 170 to 400 °C) and includes kerosene, diesel fuel (grades 1-D or 2-D), heating oil and fuel oil (grade 1 or 2), and jet fuel.[18,19] Diesel fuel is usually the fraction of petroleum that distills after kerosene. Diesel fuel can be obtained from petroleum by distillation without further processing (straight-run), by thermal cracking (broken down from larger hydrocarbons by heat), or by catalytic cracking (broken down from larger hydrocarbons by catalyst-stimulated reactions) processes. The commercial grades of diesel fuel depend on the nature of the crude oils from which they are produced, the refining practices,[20] and the blended feedstocks to achieve certain established specifications.[21]

The American Society for Testing and Materials' (ASTM) classification of diesel fuels, ASTM D 975, Standard Specification for Diesel Fuel Oils, defines three grades of diesel fuels: No. 1-D, No. 2-D, and No. 4-D. The No. 1-D grade diesel fuel comprises volatile fuel oils (from kerosene to the intermediate distillates), for use in high-speed engines with frequent and wide variations in loads and speeds and in cases where low ambient temperatures are encountered. The No. 2-D grade diesel fuel includes intermediate distillate gas oils of lower volatility, for use in high-speed engines with high loads and uniform speeds. Most automotive and truck-type diesel engines use No. 2-D fuel oil. The No. 4-D grade diesel fuel includes the more viscous distillates and blends of these distillates with residual fuel oils, for use in low- and medium-speed engines. Often the unspecified or generic diesel fuel is referred to as fuel oil No. 2;[22,33] therefore, because the type of diesel fuel originally contained in the abandoned USTs is unknown, it is assumed that fuel oil No. 2 is the mixture of concern.

Generic diesel fuel is composed chiefly of unbranched paraffins (64 percent) and other olefinic (1 to 2 percent), naphthenic, and aromatic hydrocarbons (35 percent), including alkylbenzenes and two-or three-ring aromatics.[21] The number of carbons in each constituent normally ranges from C_{10} to C_{19}.[19,21,22] The results for the TICs of VOCs and SVOCs detected in diesel-contaminated soil and groundwater samples at the site indicated that weathered diesel, as a result of environmental partitioning, transport, and transformation processes, is composed chiefly of unbranched or branched paraffins, with a low aromatic fraction of only 0 to 14 percent.[4] Due to its low aromatic content, weathered diesel may be, therefore, expected to be less toxic than its virgin analog or similar middle distillate mixtures.

Selected organic and inorganic components of a typical virgin diesel fuel No. 2 are listed in Table 1-2. The actual composition of diesel varies according to the interactions of the active ingredients, storage conditions, and the formulating ingredients added.[23] Commercial diesel fuel may also contain a variety of additives that can enhance or impart certain desirable properties to the fuel.[24] A list of diesel additives is presented in Table 1-3. The effects of chemical components, including additives, of diesel fuel are expected to be fully addressed when diesel fuel is evaluated as a mixture entity.

Table 1-2   Selected Components of Virgin Diesel Fuel

Compound	Source[a] (mg/L)	Source[b] (ppm by weight)
Organics		
Total BTEX [c]	—	1,978
Benzene	5 to 50	—
Toluene	20 to 210	—
Xylenes	20 to 1,400	—
PAHs [c]	—	2,090
2-Methylnaphthalene	1,100 to 1,200	—
Naphthalene	360 to 2,100	3,458
Phenanthrene	14 to 570	—
Pyrene	3 to 5	—
Methylphenols	—	17.6
Metals [d]		
Arsenic	<0.30	0.012 to 0.13
Cadmium	—	0.089 to 0.89
Chromium	<0.10	0.55 to 2.8
Copper	<0.03 to 0.12	—
Iron	—	3.8 to 71.0
Lead	<0.36 to 0.83	<0.49 to 2.0
Manganese	<0.02	0.29 to 6.2
Molybdenum	—	0.018 to 0.27
Nickel	<0.10	6.1 to 23.0
Phosphorous	<0.25	—
Vanadium	—	<0.06 to 0.16
Zinc	<0.07 to 0.56	1.3 to 4.8
Additional Inorganics	**(ppm)**	
Sulfur	20 to 4,800	—
Nitrogen	200 to 800	—
Chlorine	<20	—

Notes:

— Denotes no available data.

a Chevron Diesel Fuel No. 2 Composition, Substances listed in AB 2588.[23]

b The Installation Restoration Program Toxicology Guide.[21]

c BTEX = Benzene, toluene, ethylbenzene, and total xylenes; PAHs = Polynuclear aromatic hydrocarbons.

d "<" indicates concentrations below detection limits.

Table 1-3   Diesel Additive Classifications

Classification	Type	Function
Contaminant Control		
Biocides	Boron compounds, ethers of ethylene glycol, quaternary amine compounds	Inhibit growth of bacteria and fungi - prevent filter clogging
Demulsifiers and Dehazers	Surface-active materials which increase water/soil separation	Improve separation of water and prevent haze
Rust and Corrosion Inhibitors	Organic acids, amines and amine phosphates	Prevent rust and corrosion in fuel systems, pipelines, and storage facilities
Fuel Stability		
Metal Deactivators	Chelating agents	Inhibit gum formation
Oxidation Inhibitors	Alkyl amines	Minimize oxidation, gum, and precipitate formation
Dispersants	Polymeric amine surfactants	Prevent agglomeration and disperse residue
Engine Performance		
Detergents	Polyglycols, polyether amines	Prevent injector deposits Increase injector life
Dispersants	Basic nitrogen polymeric amine surfactants	Peptize injector deposits Increase filter life
Octane Improvers	Alkyl nitrates	Increase octane number
Smoke Suppressants	Overbased barium compounds	Minimize exhaust smoke
Fuel Handling		
Pour Point Depressants	Polymeric compounds	Reduce pour point and improve low-temperature fluidity properties
Cloud Point Depressants	Polymeric compounds	Reduce cloud point and improve low temperature filterability
De-icers	Low molecular weight alcohols	Reduce freezing point of small amounts of water to prevent fuel line plugging
Anti-Foam	Silicone and non-silicone surfactants	Minimize the formation of fuel foam

Source:
Diesel Fuels, Technical Publication.[24]

Physical/Chemical Properties and Environmental Fate/Transport Characteristics
The nominal physical and chemical properties that may affect the severity of the
hazard and environmental fate and transport characteristics of virgin diesel fuel are
listed in Table 1-4. These physical and chemical properties often vary depending on
different diesel components and the seasonal or geographical formulations of diesel for
various intended uses.[21] Therefore, a range, rather than a specific value, is given for
physical and chemical properties of diesel fuel. Diesel fuel is expected to have
moderate mobility and moderate persistence in most surface soils and higher

Table 1-4 Physical/Chemical and Fate/Transport Characteristics of Virgin Diesel Fuel

Characteristics	Diesel No. 2-D[a]	Fuel Oil No. 2[b]
Physical State (20°C)	Liquid	Liquid
Molecular Weight	140 g/mol	—
Color	Colorless to brown	—
Odor	Kerosene-like	—
Odor Threshold	No data	—
Density	0.870 to 0.950 g/mL (20°C)	—
Freezing/Melting Point	-48 to 18°C	—
Boiling Point	151 to >588°C	170 to 357°C (340-675°F)
Flash Point	38-74°C (No.1), 69-169°C (No. 5)	43 to 88°C (110-190°F)
Flammable Limits	0.6 to 7.5%	—
Vapor Pressure	—	2.12- 26.4 mm Hg (21°C)
Saturated Concentration in Air	no data	—
Solubility in Water	about 5 mg/L (20°C)	Negligible
Specific Gravity (water=1)	—	<0.876
Viscosity	1.152 to 1.965 cp (21°C)	2.0 to 3.6 (36°C)
Log K_{ow} (octanol-water)	3.3 to 7.06	—
Soil Adsorption Coefficient K_d	9.62E+02 to 5.5E+05	—
Henry's Law Constant	5.9E-05 to 7.4	—
Bioconcentration Factor	no data	—

Notes:
— Denotes no available data
a The Installation Restoration Program Toxicology Guide.[21]
b Material Safety Data Sheet No. 469 (Fuel Oil No. 2).[22]

persistence in subsurface soils and groundwater. Volatilization (although limited), sorption, photo-oxidation, and biodegradation are all potential fate processes,[21] causing changes in chemical composition, termed "weathering." As a result, many of the more mobile toxic chemicals originally present in diesel fuel (chemicals with high volatility and water solubility, small molecular weight, and low soil sorption capability) are removed from the environment, leaving behind chemicals of higher molecular weight, which are considered insoluble, nonvolatile, and resistant to biodegradation.[21]

Diesel fuel adsorbs strongly to soil, especially to clays. Due to its low water solubility and high soil adsorption characteristics, diesel-contaminated soil at levels below saturation is considered nonleachable.[25,26] The photo-oxidation process primarily affects aromatic compounds which may be present in low concentrations in diesel fuel. Microbiological degradation of diesel does occur, although the degradation rate is site-specific and seasonal. The n-alkanes (unbranched), n-alkylaromatic, and aromatic compounds in the C_{10} to C_{22} range are readily biodegradable under environmental conditions favorable to microbial oxidation. The branched alkanes and cycloalkanes, however, are less biodegradable than their n-alkane and aromatic analogs.[27]

Released in the subsurface in sufficient quantity, diesel fuel, as well as other LNAPLs, moves through pores and fractures of the unsaturated zone (vadose zone) downward to the groundwater surface.[28,29] During migration, these LNAPLs leave behind blobs or ganglia trapped in some of the pores and fractures, referred to as residual saturation. A three-phase system consisting of water, LNAPL, and air is then formed within the vadose zone. However, TPH-diesel is not expected to enter the vapor phase readily in the subsurface environment. Residual LNAPLs are trapped within the porous medium due to air-LNAPL and water-LNAPL interfacial tensions.[28] Since free-phase LNAPLs are lighter than water, they float on top of the water table and form two distinct plumes, each having its own average velocity.[28-30] Water table fluctuations, either regionally or locally, can emplace these LNAPLs below the water table.[27,28]

Environmental Effects of Diesel Fuel

In general, data on the environmental effects of virgin and weathered diesel are limited. The toxicity of weathered diesel fuel, as compared to its virgin analog, is primarily based on studies of marine species. To these species, diesel fuel is more toxic than kerosene, which is readily absorbed by fish and by primates through their gastrointestinal tracts.[19] The effects of diesel fuel on sea urchins are most distinct at the gastrula stage, resulting in a decrease in the larval size and asynchrony, delay of development, and an increased frequency of defects.[31] Brown shrimps and oysters in contact with the water-soluble fraction of fresh No. 2 fuel oil showed an accumulation of naphthalene and alkylnaphthalene in their tissues; however, depuration is rapid when they enter clean seawater.[32,33]

Health Effects of Diesel Fuel

Currently available human and animal toxicological data focus on the health effects of virgin diesel fuel and similar middle distillate products, such as kerosene, hydrodesulfurized kerosene, JP-5, heating oil No. 2, and hydrodesulfurized middle distillates. There are no toxicological data available on weathered diesel fuel or weathered similar middle distillate mixtures. A summary of the toxic effects of virgin diesel fuel and similar middle distillate mixtures via the oral, inhalation, and dermal routes is presented in Tables 1-5 through 1-7, respectively. In general, the physiologic responses to virgin diesel fuel and to other similar middle distillate products are similar in humans and animals.[19] Based on the results of aquatic studies on fresh and weathered oil products, the degree of health effects from weathered diesel, in the absence of known toxic volatile and semivolatile organics, as in the case of this site, are expected to be lower than those of its virgin analog.[34,35]

In humans, the immediate health effects resulting from short-term exposure to diesel involve the skin, respiratory tract, and nervous system. Diesel fuel is a moderate skin and mucous membranes irritant following prolonged or repeated direct contact. Symptoms include pain or the sensation of heat, discoloration, swelling, blistering, and development of acneiform pimples and spots. Inhalation of excessive concentrations of vapor or mist can be irritating to the respiratory system and can cause central nervous system effects such as headache, dizziness, confusion, convulsions, loss of coordination, and coma.[19,21,22,36,37] Short-term dermal and inhalation exposures to diesel fuel have also been found to induce nephropathy in humans.[21]

At low doses, diesel fuel is considered practically nontoxic to internal organs if ingested or absorbed through the skin.[36] Ingestion of a large amount of the substance may cause nausea, vomiting, and in severe cases, drowsiness to coma.[21] Chemical pneumonitis may also result when ingestion occurs and the compound is aspirated into the lungs. Severe injury to the lungs and death may follow because the substance is very difficult to remove from the lungs.[36,38,39]

Acute and systemic effects of diesel fuel and other similar middle distillate mixtures have been well studied using several animal models. In general, the health effects caused by these products are quite similar. Across the species tested, the oral LD_{50}, the lethal dose that would kill half of the dosed animals, for diesel fuel and similar middle distillate mixtures is above 5 grams per kilogram (g/kg), which indicates that these substances are not acutely toxic via ingestion.[19,40-43] Dermal studies of diesel fuel and other similar middle distillate mixtures in rabbits reported severe dermal irritation, erythema, edema, weight loss, anorexia, depression, and congested liver and kidneys.[40,41,44,45] In addition, heating oil No. 2 also caused hyperkeratosis, pyoderma, and hepatic necrosis in the 10 milliliters per kilogram (mL/kg) dose group.[41]

Inhalation of diesel or similar middle distillate mixtures produces a wide variety of physiologic responses in many animal studies, ranging from no treatment-related effects at levels around 14 ppm for kerosene or 24 milligrams per cubic meter (mg/m^3) for hydrodesulfurized middle distillate API 81-09 to effects on the neurological and respiratory systems, and the kidneys and liver.[46-51] In addition, graded airborne diesel fuel at 100 and 400 ppm was negative in a rat inhalation teratology study.[52]

Table 1-5 Toxic Effects Of Diesel Fuel And Other Middle Distillate Mixtures Via The Oral Route

Material	Species	Entry Route (Test)	Dose or Concentration	Result or Effect	Reference Number
Diesel Fuel (commercially available)	Rat	Oral (acute)	20 mL/kg	90% mortality; lethal dose to 50 percent of dosed animals (LD_{50}) = 9.0 mL/kg; 95% confidence interval (CI) = 5.6 - 14.5 mL/kg (signs: diarrhea, hemorrhagic gastroenteritis, gastrointestinal (GI) tympanites, pneumonia with abscessation)	41
No. 2 Home Heating Oil					
10% Catalytic-Cracked	Rat	Oral (acute)	20 mL/kg	90% mortality; LD_{50} = 14.5 mL/kg 95% CI = 12.3 - 17.0 mL/kg	41
30% Catalytic-Cracked	Rat	Oral (acute)	25 mL/kg	100% mortality; LD_{50} = 19.0 mL/kg 95% CI = 16.8 - 21.5 mL/kg	41
50% Catalytic-Cracked	Rat	Oral (acute)	25 mL/kg	70% mortality; LD_{50} = 21.2 mL/kg 95% CI = 18.7 - 24.9 mL/kg (toxic signs as with diesel fuel)	41
Jet Fuel A	Rat	Oral (acute)	25 mL/kg	0% mortality; LD_{50} > 25 mL/kg	41
Kerosene	Human	Oral (acute)	0.5 oz. 3 to 4 oz. 8 oz.	Lowest lethal dose Mean lethal dose Highest nonlethal dose	19
Kerosene	Human	Oral (with aspiration into lungs)	<1 mL	May cause chemical pneumonitis including central nervous system (CNS) effects	19
Kerosene	Rabbit	Oral (acute)	28 g/kg	Lethal to some animals, LD_{50}	19
	Guinea Pig	Oral (acute)	20 g/kg	LD_{50}	

Continued

Table 1-5 (Cont.)

Material	Species	Entry Route (Test)	Dose or Concentration	Result or Effect	Reference Number
Kerosene	Mouse	Oral (acute)	50 mL/kg	Intoxication with rapid respiration, congestion of renal tubules, lung surface hyperemia, liver yellow patches; death 10 hr after second equal dose on second day.	19
Kerosene	Rat	Intratracheal (acute)	800 mg/kg	Lowest lethal dose	19
Deodorized Kerosene	Rabbit Mouse	Oral (acute) Oral (acute)	64 mL/kg 64 mL/kg	Lethal to 4 out of 10 animals Lethal to 1 out of 10 animals	19
Hydrodesulfurized Middle Distillate 81-09	Rat	Oral (acute)	5 g/kg	No death. Signs: Hypoactivity, urine stained abdomen, oily fur	40
Hydro-desulfurized Kerosene	Rat	Oral (acute)	5 g/kg	Same as above	40

Table 1-6   Toxic Effects Of Diesel Fuel And Other Middle Distillate Mixtures Via The Inhalation Route

Material	Species	Entry Route (Test)	Dose or Concentration	Result or Effect	Reference Number
Diesel Fuel	Rat (female)	Inhalation (teratology)	401.5 ppm graded airborne (days 6 to 15 of gestation)	Significant decrease in food consumption; no other teratologic effects.	52
			101.8 ppm	No teratologic effects	52
Diesel Fuel No. 2 Vapor	Mouse	Inhalation (subacute neuro-toxicity)	0.204 mg/L, 5 days, 8 hr/day	Depression of inhibitory neuron and extensive depression of stimulatory neuron	47
			0.135 mg/L, 5 days, 8 hr/day	Depression of inhibitory neuron and slight depression of stimulatory neuron	47
			0.065 mg/L, 5 days, 8 hr/day	Depression of inhibitory neuron	47
Diesel Fuel	Mouse	Inhalation (dominant lethal assay)	400 ppm, 6 hr/day, 5days/wk, 8 wk	Negative	54
Aerosolized Diesel Fuel	Rat	Inhalation of aerosols (subacute)	0.5 mg/L, 2 hr	RD_{20} (respiratory rate depression in 20 percent of dosed animals) at 30 minutes of exposure transient depression in magnitude of startle reflex.	48
			3 mg/L, 2 hr	RD_{50} at 30 minutes of exposure; increased pulmonary free cells	48
			4 mg/L, 2 hr	No mortality with 9 repeated exposures	48
			6mg/L, 2 hr	6% mortality with 9 repeated exposures	48
Aerosolized Diesel Fuel	Rat (in control group)	Inhalation (subacute)	100 mg/m^3, 6 hr/day, 5 days/wk, 4 wk	No biochemical and hematological effects	88

Continued

Table 1-6 (Cont.)

Material	Species	Entry Route (Test)	Dose or Concentration	Result or Effect	Reference Number
Aerosolized Diesel Fuel	Rat	Inhalation (subacute)	1.3 mg to 6 mg, 2 hr or 6 hr, 1 to 3 times per week for 4 weeks	Initial depression in body weight, slow growth rate, and focal accumulation of pulmonary free cells	49
JP-5 (Molecular Weight [MW] 168; Boiling Point [BP] 205 to 290°C)	Rat	Inhalation of vapors (subchronic)	750 mg/m^3, 90 days	Significant increase in BUN and serum creatinine (male and female) and dose-dependent renal nephropathy and intratubular mineralized debris (male)	50
			150 mg/m^3, 90 days	Dose-dependent kidney effects in male	50
JP-5 (MW 168; BP 205 to 290°C)	Mouse	Inhalation of vapors (subchronic)	150 and 750 mg/m^3	Mild, dose-dependent reversible injury to subcellular organelles in liver	50
Kerosene	Human	Inhalation	14 ppm (0.1 mg/L)	No effects	19
Deodorized Kerosene	Rat, Dog, Cat	Inhalation (subchronic)	14 ppm, 8 hr/day, 13 wk	No discomfort in saturated vapor	19
	Rat	Inhalation of aerosols	7.4 mg/L, 6 hr/day, 4 days	Skin irritation of extremities	19
	Cat	Inhalation of aerosols	6.4 mg/L, 6 hr/day, 4 days	No effects	19
Hydrodesulfurized Middle Distillate	Rat	Inhalation (subchronic)	24 mg/m^3, 6 hr/day, 5 days/wk, 4 wks	No treatment-related effects except mild inflammatory changes in nasal tissues of males and females. Study parameters: body weight, clinical chemistry, hematology, histopathology	46
Hydrodesulfurized Kerosene	Rat	Inhalation (subchronic)	24 mg/m^3, 6 hr/day, 5 days/wk, 4 wks	No significant treatment-related effects	46

Table 1-7　Toxic Effects Of Diesel Fuel And Other Middle Distillate Mixtures Via Direct Contact

Material	Species	Entry Route (Test)	Dose or Concentration	Result or Effect	Reference Number
Diesel Fuel (commercially available)	Rabbit	Eye (primary irritation)	Undiluted, 0.1 mL/eye	Primary Eye Irritation Score (PEIS) = Non-irritating	41
No. 2 Home Heating Oil					
10% Catalytic-Cracked	Rabbit	Eye (primary irritation)	Undiluted, 0.1 mL/eye	Mildly Irritating	41
30% Catalytic-Cracked	Rabbit	Eye (primary irritation)	Undiluted, 0.1 mL/eye	Practically Non-irritating	41
50% Catalytic-Cracked	Rabbit	Eye (primary irritation)	Undiluted, 0.1 mL/eye	Practically Non-irritating	41
Jet Fuel A	Rabbit	Eye (primary irritation)	Undiluted, 0.1 mL/eye	Mildly Irritating	41
Kerosene	Rabbit	Eye (primary irritation)	Undiluted	Practically Innocuous	19
Hydrodesulfurized Middle Distillate 81-09	Rabbit	Eye (primary irritation)	Undiluted	Mildly Irritating	40
Hydrodesulfurized Kerosene	Rabbit	Eye (primary irritation)	Undiluted	Mildly Irritating	40
Diesel Fuel (commercially available)	Rabbit	Dermal (primary irritation)	Undiluted, single 0.5 mL, gauze patch	Primary Dermal Irritation Score (PDIS) = 6.81; Extremely Irritating (signs: severe erythema, edema, blistering sores)	41

Continued

Table 1-7 (Cont.)

Material	Species	Entry Route (Test)	Dose or Concentration	Result or Effect	Reference Number
No. 2 Home Heating Oil					
10% Catalytic-Cracked	Rabbit	Dermal (primary irritation)	Undiluted single 0.5 mL, gauze patch	PDIS = 3.98, Moderately Irritating	41
30% Catalytic-Cracked	Rabbit	Dermal (primary irritation)	Undiluted, single 0.5 mL, gauze patch	PDIS = 3.37; Moderately Irritating	41
50% Catalytic-Cracked	Rabbit	Dermal (primary irritation)	Undiluted, single 0.5 mL, gauze patch	PDIS = 3.83; Moderately Irritating	41
Jet Fuel A	Rabbit	Dermal (primary irritation)	Same as above	PDIS = 1.96; Mildly Irritating	41
Hydrodesulfurized Middle Distillate 81-09	Rabbit	Dermal (acute)	2 g/kg	No Death / Same	48
Hydrodesulfurized Kerosene	Rabbit	Dermal (primary irritation)	Undiluted	PDIS = 4.3 (sign = blanching, subcutaneous hemorrhage) Moderately Irritating / PDIS = 4.4	
Hydrodesulfurized Kerosene / Hydrodesulfurized Middle Distillate 81-10	Mouse	Dermal (short-term initiation/ promotion)	Undiluted	Promotional activity but not initiator / Same	60
Diesel Fuel (commercially available)	Rabbit	Dermal (subacute)	4.0 mL/kg, 24 hr/ day, 5 days/wk, 2 wks	0% mortality; Average weight (Avg. wt.) loss — 0.3 kg	41
			8.0 mL/kg, 24 hr/day, 5 days/wk, 2 wks	67% mortality; Avg. wt. loss — 0.55 kg (Signs: anorexia, depression, irritation, congested livers and kidneys)	41

Continued

Table 1-7 (Cont.)

Material	Species	Entry Route (Test)	Dose or Concentration	Result or Effect	Reference Number
No. 2 Home Heating Oil					
10% Catalytic-Cracked	Rabbit	Dermal (subacute)	2.5 mL/kg 4.0 mL/kg 10.0 mL/kg	0% mortality; Avg. wt. gain — 0.13 kg 0% mortality; Avg. wt. loss — 0.11 kg 100% mortality; Avg. wt. loss — 0.59 kg	41
30% Catalytic-Cracked	Rabbit	Dermal (subacute)	1.0 mL/kg 2.5 mL/kg 10.0 mL/kg	0% mortality; Avg. wt. loss — 0.04 kg 12.5% mortality; Avg. wt. loss — 0.05 kg 75% mortality; Avg. wt. loss — 0.46 kg	41
50% Catalytic-Cracked	Rabbit	Dermal (subacute)	1.0 mL/kg 3.0 mL/kg 10.0 mL/kg	0% mortality, Avg. wt. gain — 0.08 kg 25% mortality, Avg. wt. loss — 0.26 kg 87.5% mortality; Avg. wt. loss — 0.70 kg (Signs: as with diesel, plus hyperkeratosis, pyoderma, and hepatic necrosis at 10 mL/kg)	41
Jet Fuel A	Rabbit	Dermal (subacute)	8.0 mL/kg	75% mortality; Avg. wt. loss — 0.83 kg (signs: anorexia, depression, irritation, pale liver, and kidney)	41
Kerosene	Rabbit	Dermal (subacute)	3 mL/kg-day, 3 days	Hairloss, scaling, cracking of epidermis, no systemic toxicity	19
Kerosene	Guinea Pig	Dermal (subacute)	0.5 mL/3 days, 2 wks	Carrier for dinitrochlorobenzene, increases the reactivity, swelling, and infiltration.	19

Continued

Table 1-7 (Cont.)

Material	Species	Entry Route (Test)	Dose or Concentration	Result or Effect	Reference Number
Hydrodesulfurized Middle Distillate 81-09 / Hydrode-sulfurized Kerosene	Rabbit	Dermal (subacute)	200 mg/kg-day, 3 days/wk, 4 wks	Slight irritant / Moderate irritant	44
			1,000 mg/kg-day, 3 days/wk, 4 wks	Moderate irritant / Same	44
			2,000 mg/kg-day, 3 days/wk, 4 wks	Severe irritant / Same	44
Hydrodesulfurized Kerosene / Hydrode-sulfurized Middle Distillate	Mouse	Dermal Carcinogenesis (chronic)	0.05 mL test material diluted in toluene	Intermediate skin carcinogenicity / Same. For hydrodesulfurized kerosene, treated group: 24 mice with tumors, FEN = 41, mean latency = 76 wk; Toluene controls: 4 mice with tumors, FEN = 43, mean latency = 111 wk.	63
Petroleum Middle Distillates (BP 177 to 370°C)	Mouse	Dermal Carcinogenesis	—	Significant weak tumorigenic activity and non-neoplastic dermal changes including hyperplasia.	59

The available toxicological information in the literature on diesel fuel generally indicates negative responses in several mutagenicity tests. The American Petroleum Institute (API) has conducted a battery of three tests to evaluate the genotoxicity of diesel fuel and No. 2 fuel oil.[53-55] No. 2 fuel oil and unspecified diesel fuel were not mutagenic in the Ames Salmonella or the mouse lymphoma bioassays, and only diesel fuel was clastogenic in the bone marrow analysis.[53,55] Similarly, studies on similar middle distillate mixtures did not observe any evidence of genotoxicity.[56-58] Results of reproductive studies on diesel fuel indicated that it did not reduce the fertility of the mice.[54]

Other health effects reported for petroleum-derived materials were of dermal carcinogenic potential. In general, the dermal carcinogenic potential of petroleum-derived products is related to the PAH content of the product. It has been assumed that the middle distillates with boiling point ranges below the PAH distillation range would not have dermal carcinogenic potential.[59] Recent short-term dermal tumorigenesis and lifetime dermal carcinogenesis studies on some middle distillates found, however, that some of the petroleum-derived middle distillates are actually weak to intermediate tumor promoters.[60-66] They can also cause non-neoplastic dermal changes, including hyperplasia, which may have contributed to the tumorigenic responses. Therefore, it is likely that the dermal carcinogenic activity of the middle distillate products is the result of an epigenetic process related to skin irritation.[9,59]

Currently, the mechanism of dermal carcinogenesis and its implication in quantitative human health risk assessment are not well understood.[5,43] Researchers have found that the dermal dose-response curves from different studies of one material have significantly different slopes. Also the reproducibility of dermal carcinogenesis studies is greatly affected by the application frequency, delivery vehicles, and other physical conditions.[67,68] Due to the high level of uncertainty in this research area and their nonmutagenic property, the potential dermal carcinogenicity of diesel fuel and similar middle distillate mixtures were not considered or quantified. For purposes of the PRA, only the systemic or noncarcinogenic effects of diesel fuel as a complex mixture were addressed. An independent study performed by EPA at a later date also concluded that diesel fuel was not carcinogenic for health risk assessment purposes. Therefore, it can be assigned to the EPA weight-of-evidence Group D (not classifiable as to human carcinogenicity).[69]

Noncarcinogenic Polynuclear Aromatic Hydrocarbons (nPAHs)

Polycyclic (or polynuclear) aromatic hydrocarbons or PAHs are chemicals containing three or more fused, aromatic hydrocarbon rings. The two-ring systems (naphthalene and derivatives) and some heterocyclic systems (dibenzofuran and dibenzodioxin) can also be included as PAHs. PAH prevalence in the environment is due to both natural and anthropogenic sources. PAHs are generally found as a highly complex mixture in products resulting from incomplete combustion such as coal soot, cigarette smoke, motor vehicle exhaust, and forest fires. Seventeen PAHs are included in EPA's Hazardous Substances List, but few are well-studied.[19,70-73] For purposes of this PRA, only nPAHs detected in the soil at the site, including acenaphthene, dibenzofuran, fluoranthene, phenanthrene, and pyrene, are discussed.

Absorption of nPAHs has been demonstrated indirectly, because toxic effects have been seen after oral and inhalation exposure. nPAHs are oxidized in the liver by an enzyme, aryl hydrocarbon hydroxylase (AHH), to the epoxide, which hydrolyses to the hydroxy or dihydroxy derivative. The metabolites are the active forms of the chemicals; variations in the formation (amount, rate, products) of these metabolites account for the different effects of the various nPAHs. nPAHs also cause the synthesis of greater quantities of AHH and other drug-metabolizing enzymes; therefore, simultaneous exposure to nPAHs and other toxicants increases or decreases the toxicity of the other toxicants. A few nonmetabolic interactions also exist. nPAHs are excreted as a large variety of oxidized metabolites and conjugated metabolites, mostly through the bile into the feces.

Single oral and dermal doses of nPAHs are practically nontoxic to animals. Noncarcinogenic adverse effects associated with PAH in general and nPAH exposure have been observed in animals, but (with the exception of adverse hemolytic and dermal effects) generally not in humans. Animal studies have demonstrated that nPAHs tend to affect proliferating tissues such as the hematopoietic and lymphoid systems. The primary target of acute and chronic toxicity of nPAHs is the erythrocyte or red blood cells (hemolytic anemia), characterized by lowered hemoglobin, hematocrit, and erythrocyte values; elevated reticulocyte counts, Heinz bodies, and serum bilirubin; and fragmentation of erythrocytes. Infants appear to be particularly sensitive to the hemolytic effects of nPAHs. Varying degrees of hepatotoxicity and nephrotoxicity have also been reported in humans via oral exposure to nPAHs. There is some evidence that fluoranthene is genotoxic, but in nonhuman test systems only. The majority of the genotoxic test results for phenanthrene and pyrene are negative, although positive results were reported for each of the chemicals in at least one *in vitro* test (mutagenic in the Salmonella test). There is no evidence that acenaphthene is genotoxic, but the assays are limited to a few *in vitro* tests.

In general, nPAHs have low water solubilities. In surface water, nPAHs can volatilize, photodegrade, oxidize, biodegrade, bind to particulates, or accumulate in aquatic organisms. The transport and partitioning characteristics of nPAHs are roughly correlated to their molecular weights. Among the nPAHs detected at the site, acenaphthene and phenanthrene are considered low molecular weight compounds and fluoranthene and pyrene are considered medium molecular weight compounds. The low molecular weight nPAHs have organic carbon partition coefficient (K_{oc}) values in the range of less than 10^3 to 10^4, which indicates moderate potential to be adsorbed to organic carbon in the soil and sediments. The medium molecular weight nPAHs have K_{oc} values in the 10^4 range, with stronger soil adsorption characteristics and lower water solubility. nPAHs have also been shown to bioaccumulate in aquatic organisms and terrestrial animals and plants. Biomagnification in the aquatic food chain has not been reported because of the tendency of aquatic organisms to eliminate PAHs rapidly when they enter pollutant-free water.

EXPOSURE ASSESSMENT

Exposure is defined in EPA's risk assessment guidelines as the contact of the outer boundary of an organism with a chemical or physical agent.[12,74] The exposure assessment step estimates both the exposure concentration at exposure point of interest and the pathway-specific chronic daily intake (CDI), which is the normalized exposure rate or the mass of a chemical substance per unit body weight and unit time, for example milligrams per kilogram per day (mg/kg-day). Currently no military or civilian personnel live on or directly adjacent to the site. Because a military office complex is under construction at the site, it is unlikely that the area will have a future residential land use; however, residential RME assumptions were intentionally selected to be more conservative than actual site-specific conditions to ensure health-protective risk values. In addition, it was assumed that the highest concentrations of COCs, although detected 2 feet beneath the tank excavation at about 7 feet bgs, were ubiquitous in the site surface soils. Consequently, chronic human exposure to diesel-contaminated soil focuses on three exposure routes: (1) incidental ingestion of soil, (2) inhalation of soil particulates, and (3) dermal contact with soil. Pathway-specific CDI equations and details on the exposure factors are presented in Table 1-8 for the three exposure routes of interest.

Ingestion of Soil

In this scenario, nearby adult residents are exposed to the COCs in both outdoor soil and indoor dust via the ingestion route.[15] The incidental ingestion of soil has been long established as an important route of exposure to contaminants in the soil, particularly for children.[75] In accordance with EPA risk assessment guidelines,[15] the equation for calculating a 30-year residential exposure to soil and dust is divided into two parts: a 6-year exposure duration for young children of highest soil ingestion rate (200 milligrams per day [mg/day]) and lowest body weight (15 kilograms [kg]) and a 24-year exposure duration for older children and adults of lower soil ingestion rate (100 mg/day) and an adult body weight (70 kg). This approach will result in a time-weighted average (TWA) soil ingestion rate of 120 mg/day and body weight of 59 kg for a 30-year exposure period. These values, however, do not take into account extremely high soil ingestion activities such as the pica behavior in children, whose soil consumption rate could be between 5,000 and 8,000 mg/day,[76] and geophagia, a practice of eating soil and clay, in adults.[77]

Inhalation of Particulates

In this scenario, the on-site or nearby adult residents are exposed to the COCs in the soil via inhalation of fugitive airborne dust particulates. In the absence of actual field data, the particulate emission factor (PEF) derived from the model described by Cowherd and others[79] may be conservatively used to estimate the dust-bound contaminant concentrations in air based on existing soil concentrations.[15] The equation used to derive the PEF is as follows:[80]

$$PEF = \frac{LS \times V \times MH \times 3{,}600\,sec/hr}{A} \times \frac{1{,}000\,g/kg}{0.036 \times (1-G) \times (U_m/U_t)^3 \times F(x)}$$

where:

Parameter	Definition (Units)	Default Value[81]
LS	Width of contaminated area, meters (m)	45
V	Wind speed in the mixing zone, meters/second (m/sec)	2.25
MH	Mixing height (m)	2
A	Area of contamination, square meters (m^2)	2,025
0.036	Respirable fraction, grams/m^2-hour (g/m^2-hr)	0.036
G	Fraction of vegetative covers (unitless)	0
U_m	Mean annual wind speed (m/sec)	4.5
U_t	Threshold wind speed (m/sec)	5.4
F(x)	Function dependent on U_m/U_t (unitless)	0.9

(determined using the nomograph in a study by Cowherd and others)[79]

Dermal Contact to Soil

Dermal exposure to contaminants in soil can occur when the chemicals migrate through the semipermeable skin layer. Since many of the diesel components have a high octanol/water coefficient (K_{ow}), it is likely that there is a high risk of dermal absorption via direct skin contact.[82] The potential for chemical exposure via this route is less well studied than that for soil ingestion or soil/dust inhalation.[12,83] Unlike other exposure pathways, the exposure intakes for the dermal contact route are estimated as absorbed doses. Currently, no dermal absorption factor is available for diesel fuel as a chemical mixture. Sandmeyer reported that dermal absorption of kerosene, the main constituent of diesel fuel oil No. 1-D, is negligible, unless absorbed through injured dermal surfaces.[19] Ryan and others suggested the range of dermal absorption factors for SVOCs from the soil matrix to be 1 to 10 percent.[84] For the purposes of this PRA, the most health-protective soil dermal absorption factor for the semivolatile chemical class, or 10 percent, was assumed for diesel fuel and other nPAHs.

DOSE-RESPONSE ASSESSMENT

The dose-response assessment step of the health risk assessment process involves characterizing the relationship between the administered and/or the absorbed dose of a risk agent and the magnitude or likelihood of the critical adverse health effects. For systemic toxicants or chemicals that give rise to toxic endpoints other than cancer and gene mutations, such as diesel fuel and nPAHs, the dose-response assessment process determines a threshold value below which the adverse noncarcinogenic effects are not expected in the general population, including sensitive subgroups. The benchmark value for this threshold is the reference dose (RfD),[8,10,85] derived on the assumption that if the critical toxic effect is prevented, then all other toxic effects are prevented.

Table 1-8   Pathway-Specific Chronic Daily Intakes

Ingestion of Soil[12]

$$CDI_o = \frac{CS \times IR \times CF \times FI \times EF \times ED}{BW \times AT}$$

where:

CDI_o	=	Chronic daily intake via the ingestion route (mg/kg-day).
CS	=	Chemical concentration in soil (mg/kg).
IR	=	Ingestion rate (mg soil/day) -- assumed 120 mg/day[15]
CF	=	Conversion factor = 1E-06 kg/mg
FI	=	Fraction ingested from source (unitless) -- assumed one.
EF	=	Exposure frequency (days/year) -- assumed 350 days/year[15]
ED	=	Exposure duration (years) -- assumed 30 years[12,15,78]
BW	=	Body weight (kg) -- assumed 59 kg[15]
AT	=	Averaging time (days) -- assumed at 365 days/year for 30 years.[12]

Inhalation of Soil Particulates[80]

$$CDI_i = \frac{CS \times (1/PEF) \times IR \times FI \times EF \times ED}{BW \times AT}$$

where:

CDI_i	=	Chronic daily intake via inhalation of particulates (mg/kg-day).
CS	=	Chemical concentration in soil (mg/kg).
PEF	=	Particulate emission factor (m³/kg) -- assumed 1.9E+07 m³/kg.[15]
IR	=	Inhalation rate (m³/day) -- assumed 20 m³/day.[15]
FI	=	Fraction inhalated from source -- assumed one.
EF	=	Exposure frequency (days/year) -- assumed 350 days/year.[15]
ED	=	Exposure duration (year) -- assumed 30 years.[12,15,78]
BW	=	Body weight (kg) -- assumed. 70 kg.[12,15,78]
AT	=	Averaging time (days) -- assumed at 365 days/year for 30 years.[12]

Dermal Contact to Soil[12]

$$CDI_d = \frac{CS \times CF \times SA \times AF \times DA \times EF \times ED}{BW \times AT}$$

where:

CDI_d	=	Chronic daily intake via dermal contact (mg/kg-day).
CS	=	Chemical concentration in soil (mg/kg.)
CF	=	Conversion factor = 1E-06 kg/mg.
SA	=	Skin surface area available for contact (cm²/event) -- assumed 6,460 cm²/event.[78]
AF	=	Soil to skin adherence factor (mg/cm2) -- assumed 1 mg/cm.²[12]
DA	=	Dermal absorption factor (unitless) -- assumed 10 percent.[84]
EF	=	Exposure frequency (events/year) -- assumed 1 event/day and 150 days/year (3 times per week spent playing or gardening, except of two weeks of vacation.[15]
ED	=	Exposure duration (years) -- assumed 30 years.[12,15,78]
BW	=	Body weight (kg) -- assumed 70 kg.[12,15,78]
AT	=	Averaging time (days) -- assumed 365 days/year for 30 years.[12]

The RfD, in mg/kg-day, can be calculated using the following equation:[8,10]

$$RfD = \frac{NOAEL}{UF \times MF}$$

where:

NOAEL   =   No-Observed-Adverse-Effect Level, in mg/kg-day
UF   =   Uncertainty Factor (Ten-fold factor accounting for variation in:
-    human population [10H]
-    animal to human extrapolation [10A]
-    subchronic to chronic [10S]
-    Lowest-Observed-Adverse-Effect Level [LOAEL] tc NOAEL [10L])

MF   =   Modifying Factor (greater than 1 to 10 factor acccunting for the scientific uncertainties of the study and database).

In the risk characterization process, a comparison is often made between the NOAEL and the estimated exposure dose (EED) that includes all sources and routes of exposure involved. This ratio is called the margin of exposure (MOE).[8,10]

$$MOE = \frac{NOAEL \text{ (experimental dose)}}{EED \text{ (human dose)}}$$

If the MOE is equal to or greater than "UF x MF," the need for regulatory concern is likely to be small.

The dose-response assessment for diesel fuel was based on EPA and NRC guidelines on chemical mixtures,[9,13] which recommend using data on the mixture as an entity or on a "sufficiently similar" mixture to determine the risks associated with exposure to a complex mixture. A complex mixture is defined as a mixture that contains so many components, varying unpredictably over time and under different conditions, that any estimation of its toxicity based on its component toxicities contains too much uncertainty and error to be useful.[9] Although the preferred approach is to use *in vivo* bioassays on the mixture itself, EPA also recommends the uses of *in vivo* bioassays on similar mixtures due to the paucity of data on the mixture of concern.

A similar mixture, as defined by EPA,[7] is a mixture having the same components as the mixture of interest but in slightly different ratios, or having several common components but lacking one or more components, or having one or more additional components. The reasonable or sufficient similarity between the mixture of interest and similar mixtures can be assessed using the available physical and chemical properties and toxicological data that display similar types and degrees of toxicity. As discussed previously, the use of studies on the systemic effects of virgin diesel fuel or similar mixtures is sufficiently prudent and protective in all cases.

Toxicity Values for Diesel Fuel as a Mixture

This section presents the methodology used to derive toxicity values for diesel fuel as a mixture for the oral, inhalation, and dermal exposure routes.

Derivation of a Chronic Oral RfD for Diesel Fuel as a Mixture

Due to the limited data on the subchronic and chronic effects of diesel fuel and similar middle distillate mixtures and the innocuousness of these substances via the oral exposure route, a study by McKee and others[86] on the reproductive and subchronic effects of medium-boiling coal-derived fuel oil (from the coal liquefaction processes) was chosen as the critical study. Although the related database has large data gaps, the choice of coal-derived fuel oil as a similar mixture to diesel fuel provided the chronic oral RfD (RfD_o) representing the worst case because coal-derived fuel oil contains relatively high levels of high boiling materials (above 370 °C), such as PAHs, and more aromatic compounds that may pose a variety of toxic hazards.[87] Also, coal-derived fuel oil has been shown to be more toxic than diesel fuel when used as a positive control; it caused growth depression and decreased hemoglobin content and hematocrit values in rats exposed to 100 mg/m^3 for 4 weeks, as compared to no effects observed in the diesel fuel group.[88] Table 9 describes the critical study and equations used in deriving the chronic RfD_o for diesel fuel as a mixture. The derived chronic RfD_o is 7E-02 mg/kg-day.

Derivation of a Chronic Inhalation RfD for Diesel Fuel as a Mixture

Considerable studies have been performed to investigate the health effects via the inhalation route of diesel fuel and similar middle distillate mixtures, namely JP-5, kerosene, hydrodesulfurized kerosene, and hydrodesulfurized middle distillates. However, high-quality inhalation studies for derivation of chronic inhalation RfD (RfD_i) are still scarce. In order to select the most health-protective and representative inhalation RfD estimate, a range of RfD_is were derived using different methods and studies, such as the threshold limit value (TLV) method and the reproductive or subchronic inhalation study.[7] The most pertinent critical study selected for the RfD_i derivation purposes is the API subchronic study of whole body inhalation exposure to a hydrodesulfurized middle distillate, API 81-09.[46] This study is relevant to the derivation of an inhalation RfD for diesel fuel because API 81-09 is a semivolatile mixture. Toxicity values derived from graded airborne or vapor studies are not appropriate for use in the particulate inhalation pathway because of the lack of volatile components in weathered diesel. The test material in the API study[46] was pumped to a spraying systems atomizer, and the actual exposure concentrations were determined by standard gravimetric techniques. Table 1-10 describes the different methods used in deriving the chronic RfD_i for diesel fuel as a mixture. The results revealed that a value of 2E-03 mg/kg-day was the most health-protective estimate; therefore, it was chosen for use in the PRA.

Derivation of a Chronic Dermal RfD for Diesel Fuel as a Mixture

For the dermal route of exposure, the chronic daily intake (CDI_d) values are calculated as estimates of the absorbed doses. The dermal toxicity values must, therefore, be adjusted to absorbed doses for comparison. The chronic dermal RfD (RfD_d) for diesel fuel as a mixture was estimated using the following equation:[12]

$$RfD_d = RfD_o \times \text{Oral Absorption Factor} = 7E\text{-}02 \text{ mg/kg-day} \times 100\% = 7E\text{-}02 \text{ mg/kg-day}$$

Table 1-9   Derivation Of The Oral RfD_o For Diesel Fuel

Critical Study: McKee and others[86]

Test Material: Coal-derived fuel oil

Test Type: Reproductive and subchronic

Composition: As compared to virgin and weathered diesel fuel:

Component	Coal-Derived Fuel Oil Weight (%)	Virgin Diesel Fuel[21] Weight(%)	Site-Specific Weathered Diesel Fuel[4] Weight (%)
Saturated	36	64	>86
Aromatics	64	35	0 to 14

Boiling Range: 200 to 538 °C

General Toxicity: Dermally irritating but were not highly toxic to rats when acutely administered by oral or dermal routes.[87]

Test Method: Male and female Sprague-Dawley rats, one vehicle control group (90 females and 36 males) and 3 treatment groups (54 females and 18 males/group). Use of highly refined white oil (CAS 8012-95-1) to produce the dosing solutions. Dosages at approximately 0.02, 0.1, and 0.5 g/kg per day by gavage (dosing volume of 5 mL/kg), 5 times weekly for 13 weeks.

Results: No adverse effects in the reproductive toxicity study. Evidence of slight systemic toxicity in the subchronic toxicity study is consistent with other studies reported for diesel fuel and similar middle distillate mixtures: reduced body weights, brain weights, erythrocyte counts, hemoglobin levels, and hematocrits; and elevated liver weights and cholesterol levels, which may be associated with liver toxicity.

Subchronic NOAEL: 0.1 g/kg per day (or 100 mg/kg-day)

Critical Effects: Increasing liver weight and hematologic effects

Chronic Oral RfD_o Calculation: Using an uncertainty factor of 1,000 (10H, 10A, 10S):

$$\text{Chronic } RfD_o = 100\text{mg/kg-day} \times \frac{5 \text{ days}}{7 \text{ days}} \times \frac{1}{1,000} = 7\text{E-}02 \text{ mg/kg-day}$$

Table 1-10   Derivation Of The Inhalation RfD_i For Diesel Fuel

A. RfD_i Based On Occupational Threshold Limit Values (TLV)

Using a duration-adjustment of 4.2 (from commercial/industrial time of 8 hours/week and 5 days/week to residential time of 24 hours/day and 7 days/week) and using an uncertainty factor of 100 (10H, 10L)[82], RfC can be traditionally estimated from TLVs as follows:

$$RfC \text{ mg/m}^3 = TLV \text{ mg/m}^3 \times \frac{8 \text{ hr}}{24 \text{ hr}} \times \frac{5 \text{ days}}{7 \text{ days}} \times \frac{1}{100} = \frac{TLV \text{ mg/m}^3}{420}$$

RfD_i is then derived by multiplying the RfC by an adult inhalation rate of 20 m³/day and dividing by an average adult body weight of 70 kg.

$$RfD_i \text{ mg/kg-day} = RfC \text{ mg/m}^3 \times \frac{20 \text{ m}^3/\text{day}}{70 \text{ kg}}$$

Using a TLV of 5 mg/m³ for diesel fuel No. 2,[22] based on the TLV value for mineral oil mist, a traditional RfD_i of 3E-03 mg/kg-day was calculated for diesel fuel.

$$RfD_i \text{ mg/kg-day} = \frac{5 \text{ mg/m}^3}{420} \times \frac{20 \text{ m}^3/\text{day}}{70 \text{ kg}} = 3E\text{-}03 \text{ mg/kg-day}$$

Using a TLV of 14 ppm or 100 mg/m³ for kerosene,[19] a traditional RfD_i of 7E-02 mg/kg-day was calculated for diesel fuel.

$$RfD_i \text{ mg/kg-day} = \frac{100 \text{ mg/m}^3}{420} \times \frac{20 \text{ m}^3/\text{day}}{70 \text{ kg}} = 7E\text{-}02 \text{ mg/kg-day}$$

Table 1-10 (Cont.)   Derivation Of The Inhalation RfD_i For Diesel Fuel

B. RfD_i Based On A Reproductive Or Teratology Study

Critical Study: American Petroleum Institute[52]

Test Chemical: Graded airborne diesel fuel

Test Type: Inhalation teratology study

Test Method: Female rats (20 animals in each control and treatment group) exposed to graded airborne concentrations (0, 101.8, and 401.5 ppm) of diesel fuel on days 6 through 15 of gestation

Results: Significantly decreased food consumption in the 401.5 ppm exposure group as compared to the control. No other changes in the dams indicated an adverse compound-related effect. In addition, no evidence of compound-induced terata, variation in sex ratio, embryo toxicity, or inhibition of fetal growth and development.

Reproductive NOAEL: The experimental RfC of 401.5 ppm was converted to 2,300 mg/m^3 [19,90] as follows:

$$RfC \text{ mg/m}^3 = RfC \text{ ppm} \times \frac{MW}{24,450} \times 1,000 \text{ L/m}^3$$

$$RfC \text{ mg/m}^3 = 401.5 \text{ ppm} \times \frac{140}{24,450} \times 1,000 \text{ L/m}^3 = 2,300 \text{ mg/m}^3$$

Critical Effects: No teratological effects

Chronic inhalation RfD_i Calculation: Multiplying the RfC by an adult inhalation rate of 20 m^3/day and dividing by an average adult body weight of 70 kg and an uncertainty factor of 100 (10H, 10A):

$$RfD_i \text{ mg/kg-day} = 2.300 \text{ mg/m}^3 \times \frac{20 \text{ m}^3/\text{day}}{70 \text{ kg}} \times \frac{1}{100} = 7 \text{ mg/kg-day}$$

Table 1-10 (Cont.)   Derivation Of The Inhalation RfD$_i$ For Diesel Fuel

C. RfD$_i$ Based On Whole Body Subchronic Inhalation

Critical Study: American Petroleum Institute[46]

Test Material: Hydrodesulfurized middle distillate API 81-09, pumped to a Spraying Systems atomizer

Test Type: Whole body subchronic inhalation

Composition: As compared to virgin and weathered diesel fuel:

Component	API 81-09 Weight (%)	Virgin Diesel Fuel[21] Weight (%)	Weathered Diesel Fuel[4] Weight (%)
Paraffins (saturated)	46	64	>86
Olefins	2.5	1 to 2	—
Other Aromatics	51.5	35.0	0 to 14

Boiling range: 205 °C to 400 °C (401 °F to 752 °F)

Test Method: Twenty male and 20 female Sprague-Dawley-derived Charles River rats for each control and treatment group. Exposure period of approximately 6 hours/day, 5 days/week for four consecutive weeks. Health effect data analyses included body weight, hematological and biochemical parameters, relative organ weights, and pathologic (macroscopic and microscopic) examination of the tissues.

Results: No animals died. No exposure-related clinical signs and macroscopic findings, except subacute inflammation of the respiratory mucosal lining (rhinitis) in the 20-animal treated group (trace symptoms: 7 males and 3 females; mild symptoms: 10 males and 12 females).

Subchronic NOAEL: 24 mg/m^3

Critical effects: Mild inflammatory changes in nasal tissues or rhinitis

Table 1-10 (Cont.)   Derivation Of The Inhalation RfD_i For Diesel Fuel

C. RfD_i Based On Whole Body Subchronic Inhalation (Cont.)

Chronic Inhalation RfD_i Calculation:

1) <u>Duration- and body weight-adjustment exposure level:</u>[12,78,91] The adjusted NOAEL ($NOAEL_{ADJ}$) accounts for chronic continuous human exposure and is calculated as follows.

$$NOAEL_{ADJ}\ mg/m^3 = 24\ mg/m^3\ \times\ \frac{6\ hr}{24\ hr}\ \times\ \frac{5\ days}{7\ days}\ \times\ \frac{70\ kg}{0.38\ kg} = 789.5\ mg/m^3$$

(2) <u>Dosimetric adjustment exposure level:</u>[10] The human equivalent concentration or $NOAEL_{HEC}$ accounts for the difference in deposition of particles in various regions of the lungs across species and is estimated as follows.

$$NOAEL_{HEC}\ mg/m^3 = NOAEL_{ADJ}\ mg/m^3\ \times\ RDDR_{ER}$$

The respiratory system deposited dose ratio ($RDDR_{ER}$) is the ratio of the dose available for the entire respiratory system of the experimental animal species, $(RDD_{ER})_A$, to that of humans, $(RDD_{ER})_H$, adjusted for surface area and ventilatory volumes (Table H-1)[10].

$$RDDR_{ER} = \frac{(RDD_{ER})_A}{(RDD_{ER})_H} = 0.0095$$

The $(RDD_{ER})_A$ and $(RDD_{ER})_H$ are the function of the mass median aerodynamic diameter (MMAD) or the equivalent aerodynamic diameters (EAD). The MMAD was reported as 3.5 microns for the third week, with a geometric standard deviation ($sigma_g$) of 2.19 microns.

$$NOAEL_{HEC}\ mg/m^3 = 789.5\ mg/m^3\ \times\ 0.0095 = 7.5\ mg/m^3$$

(3) <u>Chronic inhalation RfD_i:</u> Using an uncertainty factor of 1,000 (10H, 10A, 10S):

$$Chronic\ RfD_i = 7.5\ mg/m^3\ \times\ \frac{20\ m^3/day}{70\ kg}\ \times\ \frac{1}{1000} = 2E\text{-}03\ mg/kg\text{-}day$$

Because no oral absorption factor for diesel fuel was found in the literature, professional judgment must be made in estimating the appropriate chronic RfD_d. Due to the ability of the human gastrointestinal tract to absorb small amounts of fatty acids in the C_{14} to C_{18} range and the relatively small amount of soil ingested, oral absorption efficiency for diesel fuel in the soil ingestion exposure route is considered complete or 100 percent.[89] Sandmeyer[19] also reported that the absorption of kerosene, the main constituent of diesel fuel oil No. 1-D, is rapid from the gastrointestinal tract. Table 1-11 summarizes all pathway-specific RfDs for diesel fuel and similar middle distillate mixtures, including uncertainty factors and the related critical effects.

Currently, the toxicity databases on TPH fuels for risk assessment purposes are incomplete. Draft ATSDR toxicological profiles on military fuels in general and on fuel oils in particular[92] have recently been released for public comment. The ATSDR study on fuel oils indicates that additional dose-response data are needed to determine the potential toxicity of fuel oils and diesel fuel. According to EPA, recent efforts should be taken to fill the data gaps and to study the health effects of hydrocarbon fuels, including weathered fuels.[93]

Toxicity Values for Noncarcinogenic Polycyclic Aromatic Hydrocarbons (nPAHs)
To date, there are limited toxicity values available for all nPAHs. As presented in Table 1-12, chronic RfD_o values are currently available for acenapthene (6.0E-02 mg/kg-day),[70] fluoranthene (4.0E-02 mg/kg-day),[70] and pyrene (3.0E-02 mg/kg-day).[70] To be health-protective, the RfD_o for naphthalene (4E-02 mg/kg-day)[94,95] was assumed for all other nPAHs that do not have available RfD_o, such as phenanthrene. Note that the chronic RfD_o for naphthalene was removed from the Health Effects Assessment Summary Tables (HEAST) Supplement No. 2 (November 1992)[96,97] because it is currently under review and subject to change. In addition, RfD_ds for all nPAHs were estimated using the route-to-route extrapolation as presented in the previous section and an oral absorption factor of 100 percent.[71] Table 1-12 summarizes all pathway-specific RfDs for nPAHs, including uncertainty factors and the related critical effects.

RISK CHARACTERIZATION

Risk characterization, the final step in the health risk assessment process, provides estimates of the types and magnitudes of adverse effects that a risk agent may cause in the populations of concern. To assess the noncarcinogenic risk for diesel fuel, the pathway-specific CDI is compared with the appropriate RfD to arrive at a ratio called the hazard quotient (HQ) as presented in the equation below:[12]

$$\text{Hazard Quotient} = \frac{\text{CDI}}{RfD}$$

where:

CDI   =   Chronic daily intake, averaged over the exposure duration, in mg/kg-day

RfD   =   Chronic reference dose, in mg/kg-day

The potential additivity of noncarcinogenic hazard due to exposure to multiple substances is quantified as a hazard index (HI), which is the sum of all possible chemical-specific HQs as indicated below:[12]

$$\text{Hazard Index} = \frac{CDI_1}{RfD_1} + \frac{CDI_2}{RfD_2} + ... + \frac{CDI_i}{RfD_i} = \Sigma\, HQ_i$$

To be health-protective, HIs due to various exposure pathways are assumed to be additive, as indicated in the following equation:[12]

$$\text{Total Exposure HI} = HI\,(\text{pathway}_1) + HI\,(\text{pathway}_2) + ... + HI\,(\text{pathway}_i)$$

Usually, if the HQ or HI is greater than unity, or one, meaning the exposure level exceeds the threshold RfD, a potential for adverse noncarcinogenic health effects may exist. If the HQ or HI is equal to or less than one, exposures to the COCs are not expected to result in a systemic toxic response. As the frequency of exposures exceeding the RfD increases and the size of the excess increases, the probability for adverse effects also increases. However, a clear distinction that could categorize all exposures below the RfD as "acceptable" (risk-free) and all exposures in excess of the RfD as "unacceptable" (causing adverse effects) cannot be made.[94,95] Note that HQs and HIs are not statistical probabilities; therefore, the level of concern does not increase linearly as the RfD is approached or exceeded. For regulatory purposes, an HI of 1 or less is considered to be an acceptable noncarcinogenic risk level.[11,12,16,98] If the pathway-specific or total exposure HI is greater than 1, segregation of the HI, based on the type of toxic effects or mechanisms of action, may have to be considered.[11,12]

The chemical-specific HQs resulting from exposure to diesel fuel as a mixture and nPAHs in the soil are presented in Table 1-13. Assuming that the same exposed population is consistently faced with the RME scenario for all pathways, the noncarcinogenic HI associated with exposure to diesel-contaminated soils is determined by adding the calculated HQs for each of the three exposure pathways. Results of the health risk calculation indicate that no adverse noncarcinogenic health effects are expected as a result of exposure to maximum concentrations of diesel fuel and nPAHs in the soil, because the cumulative HI of 7.3E-02 is less than unity. Also, the COCs that contribute the most to the cumulative noncarcinogenic hazard of 7.3E-02 are in diesel fuel as a mixture entity (an HI of 7.1E-02). The major pathways that drive the health risks posed by diesel-contaminated soils are dermal contact (an HQ of 5E-02), followed by ingestion (an HQ of 2E-02). Because of the uncertainty in the assessment of the dermal contact exposure route,[83] considerable conservatism was used in calculating this HQ; therefore, the actual HQ for this pathway may be much less.

Table 1-11   Oral, Inhalation, Dermal RfDs For Noncarcinogenic Effects Of Diesel As A Mixture

Mixture/ Surrogate Mixture	Oral RfD$_o$ mg/kg-day	Inhalation RfD$_i$ mg/kg-day	Dermal RfD$_d$ mg/kg-day	Confidence Level	Critical Effects	Basis	Reference Number	Uncertainty and Modifying Factors (a)
Coal-Derived Fuel Oil (b)	7E-02	—	—	Medium	Increased liver weight, hematologic effects	Gavage	86	1000(H,A,S)/1
Diesel No. 2 (Based on mineral oil mist)	—	3E-03	—	Low	Occupational TLV	Inhalation	22	100(H,L)/1
Kerosene	—	7E-02	—	Medium	Occupational TLV	Inhalation	19	100(H,L)/1
Graded Airborne Diesel	—	7E+00	—	Medium	Teratologic effects	Inhalation	52	100(H,A)/1
Hydrodesulfurized Middle Distillate (b)	—	2E-03	—	Medium	Mild rhinitis	Inhalation	46	1000(H,A,S)/1
Coal-Derived Fuel Oil (b)	—	—	7E-02	Low	Increased liver weight, hematologic effects	Gavage	86	Route-to-route extrapolation

Notes:

— Indicates Not Applicable

(a) Uncertainty factors (UF) are ten-fold factors used to represent combined H,A,S, and L extrapolations:

    H=Variation in human sensitity

    A=Animal to human extrapolation

    S=Extrapolation from subchronic to chronic No Observable Adverse Effect Level (NOAEL)

    L=Extrapolation from Lowest Observable Adverse Effect Level (LOAEL) to NOAEL

Modifying factors (MF), varying from 1 to 10, are applied to reflect professional judgement regarding additional uncertainties in the study and the database.

(b) RfD from this study was selected for quantitative risk assessment purposes.

Table 1-12   Oral, Inhalation, Dermal RfDs For Noncarcinogenic Effects Of Detected nPAHs In Soil

Chemical	Oral RfD$_o$ mg/kg-day	Inhalation RfD$_i$ mg/kg-day	Dermal RfD$_d$ mg/kg-day	Confidence Level	Critical Effects	Basis	Reference Number	Uncertainty and Modifying Factors (a)
Acenaphthene (b,c)	6.0E-02	6.0E-02	6.0E-02	Low	Liver toxicity	Gavage	70	3000(A,H,S,D)/1
Fluoranthene (b,c)	4.0E-02	4.0E-02	4.0E-02	Low	Kidney, liver, and blood	Gavage	70	3000(A,H,S,D)/1
Phenanthrene (b,c)	4.0E-02	4.0E-02	4.0E-02	(d)	(d)	(d)	(d)	(d)
Pyrene (b,c)	3.0E-02	3.0E-02	3.0E-02	Low	Kidney effects	Gavage	70	3000(A,H,S,D)/1

Notes:

RfD--Reference Dose

(a) Uncertainty factors (UF), valued 10 or less each: H =Variation in human sensitivity; A= Animal to human extrapolation; S=Extrapolation from subchronic to chronic; D=Lack of toxicity studies in a second species and developmental or reproductive studies; L =Extrapolation from Lowest-Observed-Adverse-Effects-Level (LOAEL) to No-Observed-Adverse-Effects-Level (NOAEL).
   Modifying factors (MF), varying from 1 to 10, are applied to reflect professional judgement regarding additional uncertainties in the study and the entire database.

(b) Inhalation RfD$_i$ was conservatively assumed to be equal to oral RfD$_o$.

(c) Oral absorption factor was assumed to be 100 percent in deriving dermal RfD$_d$ from oral RfD$_o$.

(d) Toxicity value is conservatively assumed to be that of naphthalene.[95-97]

Table 1-13 Noncarcinogenic Hazard Posed By Diesel-Contaminated Soils At SB-01

Chemical	Soil Concentration (CS) mg/kg	Soil/Dust Ingestion						Dermal Contact to Soil/Dust			Chemical-Specific HI (e)
		CDI_o (a) mg/kg-day	RfD_o mg/kg-day	HQ (b)	CDI_i (c) mg/kg-day	RfD_i mg/kg-day	HQ (b)	CDI_d (d) Mg/kg-day	RfD_d mg/kg-day	HQ (b)	
TPH As Diesel	800	1.6E-03	7E-02	2.2E-02	1.2E-05	2E-03	5.8E-03	3.0E-03	7E-02	4.3E-02	7.1E-02
Acenaphthene	1.1	2.1E-06	6E-02	3.6E-05	1.6E-08	6E-02	2.6E-07	4.2E-06	6E-02	7.0E-05	1.1E-04
Fluoranthene	5.5	1.1E-05	4E-02	2.7E-04	7.9E-08	4E-02	2.0E-06	2.1E-05	4E-02	5.2E-04	7.9E-04
Phenanthrene	5.5	1.1E-05	4E-02	2.7E-04	7.9E-08	4E-02	2.0E-06	2.1E-05	4E-02	5.2E-04	7.9E-04
Pyrene	2.1	4.1E-06	3E-02	1.4E-04	3.0E-08	3E-02	1.0E-06	8.0E-06	3E-02	2.7E-04	4.0E-04
Total				2.3E-02			5.8E-03			4.5E-02	7.3E-02

Notes: CDI-Chronic Daily Intake   RfD-Reference Dose   HQ-Hazard Quotient   HI-Hazard Index

(a) Oral CDI_o=(CSxIRxCFxFIxEFxED)/(BWxAT), as administered dose (mg/kg-day), where:

    CS=Chemical Concentration in Soil, in mg/kg

        IR=Ingestion Rate, in mg/day=   120

        CF=Conversion Factor, in kg/mg=   1E-06

        FI=Fraction Ingested from Source=   1

        EF=Exposure Frequency, in days/year=   350

        ED=Exposure Duration, in years=   350

        BW=Body Weight, in kg=   59

        AT=AveragingTime, in days=   10,950

(b)Pathway-specific HQ=CDI/RfD

(c) Inhalation CDI_i=(CSx[1/PEF]xIRxFIxEFxED)/(BWxAT), as administered dose (mg/kg-day), where:

    CS=Chemical Concentration in Soil, in mg/kg

    PEF=Particulate Emission Factor, m³/kg=   1.9E+07

    IR=Inhalation Rate, in m³/day=   20

    FI=Fraction Inhaled from Source=   1

    EF=Exposure Frequency, in days/year=   350

    ED=Exposure Duration, in years=   30

    BW=Body Weight, in kg=   70

    AT=Averaging Time, in days=   10,950

(d) Dermal CDI_d=(CSxCFxSAxAFXDAxEFxED)/(BW X AT), as absorbed dose (mg/kg-day), where:

    CS=Chemical Concentration in Soil, in mg/kg

    CF=Conversion Factor, kg/mg=   1.0E-06

    SA=Skin Surface Area, in cm²/event=   6,460

    AF=Soil-to-Skin Adherence Factor, mg/cm²=   1

    DA=Dermal Absorption Factor, unitless=   0.10

    EF=Exposure Frequency, in events/year=   150

    ED=Exposure Duration, in years=   30

    BW=Body Weight, in kg=   70

    AT=Averaging Time, in days=   10,950

(e) An overall cumulative HI of 0.07 indicates that no adverse noncarcinogenic effects are expected as a result of exposure to diesel fuels and nPAHs.

For diesel fuel as a mixture, the MOE, using the NOAEL of 100 mg/kg-day and the EED via three exposure pathways of 4.6E-03 mg/kg-day, is illustrated below:

$$\text{MOE} = \frac{\text{NOAEL (experimental dose)}}{\text{EED (human dose)}} = \frac{100 \text{ mg/kg-day}}{4.6\text{E-}03 \text{ mg/kg-day}} = 21,740$$

Since the MOE is above the UF of 1,000, the oral RfD_o, being the highest RfD, is expected to be health protective.

UNCERTAINTIES

The cumulative HI calculated above presents a conservative RME point estimate of the health risk posed by diesel-contaminated soil under a residential land-use setting. The actual risk posed by the diesel-contaminated soil is believed to be nonexistent due to the lack of complete exposure pathways at the site. Assuming that three complete exposure pathways exist, assumptions regarding site conditions (land use, chemical distribution), RME exposure levels, toxicity values, and risk calculations are considerably conservative. The use of the maximum diesel concentration detected in soils provides the initial measure for the health-protective approach in the PRA process. The RME estimated for each exposure pathway was based on many health-protective assumptions and upperbound exposure parameter values, such as the upper 90th or 95th confidence limit on contact rate, exposure frequency, and exposure duration.

Released into the environment, some of the components of virgin diesel and similar mixtures can partition into different environmental compartments, such as air, in the case of aromatics and volatiles, and groundwater, in the case of the water-soluble fractions. A considerable amount may also degrade or transform at different rates in the environment. The use of toxicological data on similar virgin mixtures that contain considerable amounts of BTEX and high boiling materials, such as PAHs, is very conservative when applied to weathered diesel fuel, as evidenced by the absence of BTEX and low levels of PAHs in soil samples from the site. A decreasing factor for the actual health risk is expected, based on aquatic toxicity of fresh and weathered petroleum oil.[34,35] According to Bobra and others,[34] weathered crude oil had a marked decrease in water solubility and thus was less toxic than crude oil in a way that it was impossible to form a lethal aqueous solution of weathered crude oil.

The summation of HQs across all pathways to arrive at an overall cumulative HI for the site is the last health protective measure in the PRA process. In reality, the same population of interest may not experience the RME level consistently for more than one pathway over the same period of time. One individual may face the RME through one pathway, and another individual may face the RME through a different pathway. The total exposure and HI presented in the PRA, therefore, represent the potential reasonable maximum risk to the population of concern should direct contact to the contaminated soil exist.

HEALTH-BASED CLEANUP GOAL (HBG) FOR DIESEL FUEL IN SOIL

From the results of the PRA, only HBG for diesel fuel in the soils (HBG_{soil}), based on the level of concern associated with noncarcinogenic effects, was derived because diesel fuel as a mixture was found to pose the most to the hazard at the site. According to the National Contingency Plan,[98] noncarcinogenic effect-based HBGs are set so that exposures to those levels will present no appreciable risk of significant adverse effects to the human receptors, namely a cumulative HI of one or unity.[16] Although the HBGs could be developed using the inverse intake algorithms discussed in the Exposure Assessment Section[80], a comparable and simpler approach is presented below:[99]

$$\frac{\text{Preliminary Risk}}{\text{Preliminary Concentration}} = \frac{\text{Target Risk}}{HBG_{soil}}$$

Therefore,

$$HBG_{soil} = \frac{\text{Preliminary Concentration x Target Risk}}{\text{Preliminary Risk}}$$

The preliminary risk is the hazard index posed by diesel fuel via the three exposure pathways previously calculated, that is, an HI of 7.1E-02. The preliminary concentration is the soil concentration of diesel used in the PRA, namely 800 mg/kg, and the target risk is an acceptable cumulative HI of 1. Both of the above equations take into account the additivity of the risk via all exposure pathways considered.

The resultant HBG_{soil} for TPH-diesel in the soil, therefore, is shown in the following equation:

$$HBG_{soil} = \frac{800 \text{ mg/kg x 1}}{0.071} = 11{,}000 \text{ mg/kg or 11 g/kg}$$

Since none of the TPH-diesel levels detected in soil samples from the site approach the 11,000 mg/kg cleanup goal, the potential hazard from existing diesel-contaminated soil appears to be minimal. The 11,000 mg/kg cleanup goal was derived on health protective assumptions regarding land uses, exposure levels, toxicity, and risk assessment methodology. HBG_{soil} values based on site-specific conditions may in fact be higher than 11,000 mg/kg. The HBGs derived herein should be used as reference only for environmental work. They are not construed to be absolute cleanup standards since they were derived based upon risk calculations with certain inherent uncertainties. As additional information on diesel fuel becomes available, the results presented in this paper should be updated accordingly.

GROUNDWATER PROTECTIVENESS OF HBG$_{soil}$ FOR DIESEL FUEL

To determine whether the HBG$_{soil}$ of 11,000 mg/kg for diesel-contaminated soil based on direct contact is protective of the underlying aquifer at the site, a health-based level for diesel fuel in soils based on potential migration from soil to groundwater (C_s) in mg/kg, must be derived and compared with the value of 11,000 mg/kg. Estimating the HBG for diesel-contaminated soil for groundwater protection purposes consists of three steps:

(1) Estimation of the health-based cleanup goal for diesel fuel in groundwater (HBG$_{gw}$), in milligrams per liter (mg/L).

(2) Estimation of the leachate concentration (C_p), in mg/L, using the HBG$_{gw}$ and the Summer's leachate model[100] or the generic California Regional Water Quality Control Board's (CRWQCB) environmental attenuation factor (EAF).[101] In the absence of site-specific data, the use of a default EAF of 100 is recommended to conservatively represent the average environmental attenuation of chemicals. This value is supported by two studies conducted by the Battelle Pacific Northwest Laboratories[102] and EPA.[103] CRWQCB describes the average EAF of chemicals to be the case for groundwater at significant depth (greater than 30 feet bgs) and for soils containing appreciable and continuous clay or silty clay strata between the base of the site and groundwater.[101]

(3) Estimation of the C_s, in mg/kg, by multiplying the leachate concentration C_p by the soil-water equilibrium partition coefficient (K_d), in liters per kilogram (L/kg).[100] The K_d for the diesel fuel was calculated by multiplying the K_{oc} by the fraction of organic carbon content of the soil (f_{oc}), assumed to be 0.02, or 2 percent.[80] Since diesel is a complex hydrocarbon mixture, it has a range of K_{oc} values for all components, which are partly dependent on the size of the hydrocarbon chain. The range of K_{oc} values for all components in diesel fuel reported in the literature is 9.62E+02 L/kg (naphthalene $C_{10}H_8$) to 5.5E+05 L/kg (benzo[a]pyrene $C_{20}H_{10}$).[21] Because phenanthrene ($C_{14}H_{10}$) was detected at the highest concentration among all nPAHs in the soil and it is the longest hydrocarbon chain detected in groundwater, the K_{oc} value for phenanthrene (1.4E+04 L/kg) was used as the appropriate surrogate for weathered diesel. Hydrocarbons and PAHs with carbon chains longer than phenanthrene are not of concern because they adsorb strongly to the soil and are practically insoluble in water.[71] Thus, they are immobile in the soil/water system.

Table 1-14 presents the equations and results for calculating C_s for diesel fuel as a mixture. Because C_s is higher than HBG$_{soil}$, the results indicate that a HBG$_{soil}$ of 11,000 mg/kg for diesel fuel based on direct contact is protective of the underlying aquifer should the groundwater be used for potable purposes.

Table 1-14   Estimation Of The Cs For Diesel Fuel For Groundwater Protection Purposes

Estimation of a health-based level for diesel fuel in soils based on potential migration from soil to groundwater (C_s) consists of three steps.

(1)   Estimation of the health-based cleanup goal for groundwater (HBG_{gw}), in milligrams per liter (mg/L).

$$HBG_{gw} \; (mg/L) = \frac{RfD \; \times \; BW \; \times \; AT}{WI \; \times \; FI \; \times \; EF \; \times \; ED} = 2.6 \; (mg/L)$$

where:

RfD	=	Oral reference dose for diesel, 7E-02 mg/kg-day.
BW	=	Lifetime average body weight, 70 kg.[12,15,78]
AT	=	Averaging time, 365 days/year for 30 years or 10,950 days.[12,15]
WI	=	Average daily water intake, 2 L/day.[12,15,78]
FI	=	Fraction ingested from source, assumed one.[12]
EF	=	Exposure frequency, 350 days/year.[15]
ED	=	Exposure duration, 30 years.[12,15]

Note that the 1991 HDOH's recommended cleanup goal for diesel-contaminated groundwater is 10 mg/L.[2]

(2)   Estimation of the leachate concentration (C_p), in mg/L, using the HBG_{gw} and the Summer's leachate model[100] or the generic California Regional Water Quality Control Board's (CRWQCB) environmental attenuation factor (EAF):[101]

$$Cp \; (mg/L) = HBG_{gw} \; (mg/L) \; \times \; 100 = 2.6 \; mg/L \; \times \; 100 = 260 \; mg/L$$

(3)   Estimation of the C_s, in mg/kg, by multiplying the leachate concentration C_p by the soil-water equilibrium partition coefficient (K_d), in liters per kilogram (L/kg):[100]

$$C_s \; (mg/kg) = K_d \; (L/kg) \; \times \; C_p \; (mg/L)$$

$$K_d = K_{oc} \; \times \; f_{oc}$$

Using the K_{oc} value for phenanthrene (1.4E+04 L/kg) as the appropriate surrogate for weathered diesel fuel and a f_{oc} value of 0.02,[80] C_s is estimated as:

$$C_s \; (mg/kg) = 260 \; mg/L \; \times \; 1.4E+04 \; L/kg \; \times \; 0.02 = 72,800 \; mg/kg$$

The results indicate that a HBG_{soil} of 11,000 mg/kg for diesel fuel based on direct contact is protective of the underlying aquifer should the groundwater be used for potable purposes.

CONCLUSIONS

Although TPH-diesel concentrations in soils at the site exceeded the HDOH's 1991 recommended cleanup goals for diesel-contaminated soil, the absence of complete exposure pathways indicated that the contamination at the site did not pose an immediate threat to human health or to the environment. This was further substantiated by the results of the PRA, which demonstrated that the noncarcinogenic risk posed by the maximum diesel fuel and nPAH concentrations detected at the site was still below the EPA's acceptable noncarcinogenic risk level. An HBG_{soil} of 11,000 mg/kg for diesel fuel as a mixture was calculated using the results of the PRA. Soil TPH-diesel levels at the site were much lower than this level, verifying that the existing diesel-contaminated soil posed insignificant risk. As a result, a no-action remedial alternative was recommended for the diesel-contaminated soils and the proposal concurred by the HDOH. In 1993, monitoring groundwater wells were installed at the site and a risk assessment of the diesel-contaminated groundwater was performed. The ground-water modeling and risk assessment results indicated that a remediation of groundwater at this site appeared unwarranted. It should be noted that toxicity values for diesel fuel as a complex mixture were derived for quantitative risk assessment purposes; assumptions used in this derivation, while health protective, are subject to interpretation.

ACKNOWLEDGMENT

Part of this manuscript was reprinted with the written permission of the Hazardous Materials Control Resources Institute from the paper "Derivation of Toxicity Values for Diesel Fuel and Related Risk Assessment" by Pham-Mahini and others published in the HMC/Superfund'92 Proceedings. The author wishes to express her gratitude to Dr. Paul Kostecki for his support and to Thorsten Anderson (Project Chemist), Bill Wade (Editorial Editor), Daniel Chow (QC Reviewer), Jill S. Yamada (Project Manager), Eric S. Morton (Technical Editor) Harry V. Ellis III (Technical Editor) and Tamlyn Oliver (Publications Director) for their contribution to the manuscript.

REFERENCES

1.   Fewell Geotechnical Engineering. Ltd., Underground Storage Tank Site Assessment Report, U.S. Army Reserve Center, Fort Shafter, Oahu, Hawaii, Prepared for RHS Lee, Inc., 1991.

2.   Hawaii Department of Health (HDOH), Soil and Groundwater Clean-up Goals for Typical Petroleum Constituents, Solid and Hazardous Waste Branch, Underground Storage Tank Section, 1991.

3.   Clayton Environmental Consultants, Soil Excavation at the U.S. Army Reserve Center, Fort Shafter Flats, Phase II, Site No. 2, Honolulu, Oahu, Hawaii, Prepared for the U.S. Army Corps of Engineers District, Honolulu, 1991.

4.   PRC Environmental Management, Inc., Underground Storage Tank Characterization at the U.S. Army Reserve Center Site, Fort Shafter Flats, Hawaii, February 1992.

5.   Dourson, Michael, U.S. EPA, Environmental Criteria and Assessment Office, Cincinnati, Ohio, personal communication with Mahini, Xuannga, September 18, 1991.

6.   Velazquez, S., U.S. EPA Toxicologist, Environmental Criteria and Assessment Office, Cincinnati, personal communication with Mahini, Xuannga, September 30, 1991.

7.   U.S. Environmental Protection Agency, Guidelines for the Health Risk Assessment of Chemical Mixtures, *Federal Register*, Volume 51: 34014, 1986.

8.   U.S. Environmental Protection Agency, Integrated Risk Information System (IRIS) Background Document 1 - Reference Dose (RfD): Description and Use in Health Risk Assessments, 1988.

9.   U.S. Environmental Protection Agency, Technical Support Document on Risk Assessment of Chemical Mixtures, Office of Health and Environmental Assessment, Cincinnati, Ohio 45268, EPA/600/8-90/064, 1988.

10.   U.S. Environmental Protection Agency, Interim Methods for Development of Inhalation Reference Doses, EPA/600/8-88/066F, 1989.

11.   U.S. Environmental Protection Agency, Risk Assessment Guidance for Superfund - Human Health Risk Assessment, U.S. EPA Region IX Recommendations, Interim Final, 1989.

12.   U.S. Environmental Protection Agency, Risk Assessment Guidance for Superfund - Human Health Evaluation Manual (Part A), Interim Final, Office of Emergency and Remedial Response, EPA/540/1-89/002, 1989.

13.   National Research Council (NRC), *Complex Mixtures: Methods for In Vivo Toxicity Testing*, National Academy Press, Washington, D.C., 1988.

14.   Schoeny, R. S., Creative approaches in the study of complex mixtures: Evaluating comparative potencies, in *Principles and Practices for Petroleum Contaminated Soils*, (Calabrese, Edward J. and Kostecki, Paul T., eds.), Lewis Publishers, 25, 591, 1993.

15.   U.S. Environmental Protection Agency, Risk Assessment Guidance for Superfund, Volume I: Human Health Evaluation Manual - Supplemental Guidance "Standard Default Exposure Factors", Interim Final, OSWER Directive 9285.6-03, 1991.

16. U.S. Environmental Protection Agency, Role of the Baseline Risk Assessment in Superfund Remedy Selection Decisions, Memorandum from Assistant Administrator Don R. Clay, OSWER Directive 9355.0-30, April 22, 1991.

17. U.S. Council on Environmental Quality, Executive Office of the President, *Risk Analysis: A Guide to Principles and Methods for Analyzing Health and Environmental Risks*, 1989.

18. Williams, J., Refining Department, American Petroleum Institute, personal communication with Mahini, Xuannga, October, 1991.

19. Sandmeyer, E. E., Aromatic hydrocarbons, in *Patty's Industrial Hygiene and Toxicology*, Chapter 47, Volume 2B, 3rd Ed, 1981.

20. Saary, Z., Chevron U.S.A., personal communication with Mahini, Xuannga, September 25, 1991.

21. U.S. Department of Energy (DOE), The Installation Restoration Program Toxicology Guide, Health and Safety Research Division, Oak Ridge National Laboratory, 1989.

22. Genium Publishing Corporation, Material Safety Data Sheet No. 469 (Fuel Oil No. 2), 1981.

23. Chevron Research Company, Chevron Diesel Fuel No. 2 Composition, Substances Listed in AB 2588, 1991.

24. Chevron Research Company, Diesel Fuels, Technical Publication, Richmond, California, 1988.

25. Mackay, D., The chemistry and modeling of soil contamination with petroleum, in *Soils Contaminated by Petroleum: Environmental and Public Health Effects*, John Wiley and Sons, New York, 5, 1988.

26. Henderson, U. V., An overview of public and environmental health and safety, in *Soils Contaminated by Petroleum: Environmental and Public Health Effects*, John Wiley and Sons, New York, 169, 1988.

27. Dragun, J., Microbial degradation of petroleum products in soil, in *Soils Contaminated by Petroleum: Environmental and Public Health Effects*, John Wiley and Sons, New York, 289, 1988.

28. Hunt, J. R., Shar, N., and Udell, K., Nonaqueous phase liquid transport and cleanup: 1. Analysis of mechanisms, *Water Resources Research*, 24 (8), 1,247, 1988.

29. U.S. Environmental Protection Agency, Handbook: Ground Water, Volume II: Methodology, EPA/625/6-90/016b, July, 1991.

30. U.S. Environmental Protection Agency, Superfund Exposure Assessment Manual, Office of Remedial Response, EPA/540/1-88/001, 1988.

31. Vashchenko, M. A., Durkina, V. B., and Gnezdilova, S. M., Effect of diesel fuel hydrocarbons and cadmium on the development of sea urchin progeny, Language: Russian, *Ontogenez*, 19 (1), 82, 1988.

32. American Petroleum Institute, Effects of Oil and Chemically Dispersed Oil on Selected Marine Biota, A Laboratory Study, Report No. 26-60050, *API Medical Research Publication* 4191: 87, 1973.

33.   American Petroleum Institute, The Effects of Oil on Estuarine Animals: Toxicity, Uptake, and Depuration, Respiration, Report No. 26-60030, 1974.

34.   Bobra, A. M., Shiu, W. Y., and Mackay, D., Acute toxicity of fresh and weathered crude oils to Daghnia Magna, *Chemosphere*, 12 (9-10), 1137, 1983.

35.   Hertzberg, R., Environmental Criteria and Assessment Office, Cincinnati, Ohio, personal communication with Mahini, Xuannga, October 1, 1991.

36.   Chevron Environmental Health Center, Material Safety Data Sheet No. CPS272102 for Chevron Diesel Fuel No. 2, Revision No. 10, 1990.

37.   Porter, H. O., Aviators intoxicated by inhalation of JP-5 fuel vapors, *Aviation Space and Environmental Medicine*, 61 (7), 654, 1990.

38.   Lee, C. H., Chiang, Y. C., Lan, R. S., Tsai, Y. H., and Wang, W. J., Aspiration pneumonia following diesel-oil siphonage-analysis of 12 cases, *Chang Keng I Hsueh Chang Gung Medical Journal*, 11 (3), 180, 1988.

39.   Vertkin, I. I., and Platunov, S. K., A case of pneumonia caused by aspiration of diesel fuel, Language: Russian, *Gigiena Truda i Professionalnye Zabolevaniia*, (3), 41, 1989.

40.   American Petroleum Institute, Acute Toxicity Studies of Hydrodesulfurized Middle Distillate, [API] Sample 81-09, Report No. 30-32347, *API Medical Research Publication*, 1982.

41.   American Petroleum Institute, The Acute Toxicology of Selected Petroleum Hydrocarbons, Report No. 30-31530, *API Publication: Applied Toxicolology of Petroleum Hydrocarbons*, 1984.

42.   Parker, G. A., Bogo, V., and Young, R. W., Acute toxicity of petroleum and shale-derived distillate fuel, marine: Light microscopic, hematologic, and serum chemistry studies, *Fundamental and Applied Toxicology*, 7 (1), 101, 1986.

43.   McKee, R., Exxon Biomedical Sciences, New Jersey, personal communication with Mahini, Xuannga, September 26, 1991.

44.   American Petroleum Institute, Twenty-eight-day Dermal Toxicity Study in the Rabbit of Hydrodesulfurized Middle Distillate, [API] Sample 81-09, Report No. 30-32298, *API Medical Research Publication*, 1983.

45.   American Petroleum Institute, Project No. 1443/Acute Toxicity Tests API 79-6 Diesel Fuel (Marketplace Sample), Report No. 27-32817, *API Medical Research Publication*, 1980.

46.   American Petroleum Institute, Four-week Subchronic Inhalation Toxicity Study in Rats, API 81-07, Hydrodesulfurized Kerosene (Petroleum) (CAS 64742-81-0); API 81-09 and API 81-10, Hydrodesulfurized Middle Distillate (Petroleum) (CAS 64742-80-9), Final Report, Report No. 33-32724, *API Health and Environmental Science Department Report*, 1986.

47.   American Petroleum Institute, Depressant Effects Associated With The Inhalation of Uncombusted Diesel Vapor, Report No. 30-32078, *API Publication: Applied Toxicology of Petroleum Hydrocarbons*, 1984.

48.   American Petroleum Institute, [A Series of Nine] Inhalation Exposures of [Sprague-Dawley] Rats to Aerosolized Diesel Fuel, Report No. 30-31531, *API Medical Research Publication*, 13, 1982.

49.   Dalbey, W., Henry, M., Holmberg, R., Moneyhun, J., Schmoyer, R., and Lock, S., Role of exposure parameters in toxicity of aerosolized diesel fuel in the rat, *Journal of Applied Toxicology*, 7 (4), 265, 1987.

50.   American Petroleum Institute, Toxicology of Mixed Distillate and High-Energy Synthetic Fuels, Report No. 32-30415, *API Publication*, 1984.

51.   American Petroleum Institute, Pathologic Findings in Laboratory Animals Exposed to Hydrocarbons Fuels of Military Interest, Report No. 32-30416, *API Publication*, 121, 1984.

52.   American Petroleum Institute, Teratology Study in Rats, Diesel Fuel, Final Report, Report No. 27-32174, A*PI Medical Research Publication*, 1979.

53.   American Petroleum Institute, Mutagenicity Evaluation of Diesel Fuel, Report No. 26-60102, 1978.

54.   American Petroleum Institute, Mutagenicity Evaluation of Diesel Fuel in the Mouse Dominant Lethal Assay, Final Report, Report No. 28-31346, 1980.

55.   American Petroleum Institute, In-vitro and In-vivo Mutagenicity Studies, No. 2, Home Heating Oil, Financial Report, Report No. 27-30140, *API Medical Research Publication*, 1979.

56.   Henderson, T. R., Li, A. P., Royer, R. E., and Clark, C. R., Increased cytotoxicity and mutagenicity of diesel fuel after reaction with NO2, *Environmental Mutagenesis*, 3, 211, 1981.

57.   Lebowitz, H., Brusick, D., Matheson, D., Jagannath, D. R., Reed, M., Goode, S., and Roy, G., Commonly used fuels and solvents evaluated in a battery of short-term bioassays, *Environmental Mutagenesis*, 1, 172, 1979.

58.   Zeiger, E., Anderson, B., Haworth, S., Lawlor, T., Mortelmans, K., and Speck, W., Salmonella mutaginicity tests - 3. Results from the testing of 255 chemicals, *Environmental Mutagenesis*, 9 (Supplement 9), 110, 1987.

59.   Biles R. W., McKee, R. H., Lewis, S. C., Scala, R. A., and DePass, L. R., Dermal carcinogenic activity of petroleum-derived middle distillate fuels, *Toxicology*, 53 (2-3), 301, 1988.

60.   American Petroleum Institute, Short-term Dermal Tumorigenesis Study of Selected Petroleum Hydrocarbon in Male CD-1 Mice, Initiation and Promotion Phases of Dermal Tumorigenesis, Report No, 36-32643, *API Health and Environmental Science Department Report*, 1989.

61.   Clark, C. R., Walter, M. K., Ferguson, P. W., and Katchen, M., Comparative dermal carcinogenesis of shale and petroleum-derived distillates, *Toxicology and Industrial Health*, 4 (1), 11, 1988.

62.   Rothman, N. and Emmett, E. A., The carcinogenic potential of selected petroleum-derived products, State of the Art Reviews, *Occupational Medicine*, 3 (3), 475, 1988.

63.   American Petroleum Institute, Lifetime Dermal Carcinogenesis Bioassay of Refinery Streams in C_3H/HeJ Mice (AP-135R), Report No. 36-31364, *API Health and Environmental Science Department Report*, 1989.

64.   McKee, R. H., Plutnick, R. T., and Przygoda, R. T., The carcinogenic initiating and promoting properties of a lightly refined paraffinic oil (A middle distillate product), *Fundamental and Applied Toxicology*, 12 (4), 748, 1989.

65.   National Toxicology Program, Toxicology and Carcinogenesis Studies of Marine Diesel Fuel and JP-5 Navy Fuel (CAS No. 8008-20-6) in B6C3F1 Mice (Dermal Studies), NTP Technical Report Series No. 310, 1986.

66.   Schultz, T. W., Epler, J. L., Witschi, H., Fry, R. J. M., Smith, L. H., Rao, T. K., Haschek, W. M., Larimer, F. M., Holland, J. M., and Dumont, J. N., Health Effects Research in Oil Shale Development, Oak Ridge National Laboratory Report, ORNL/TM-8034, 1981.

67.   McKee, R. H., Nicolich, M. J., Scala, R. A., and Lewis, S. C., Estimation of epidermal carcinogenic potency, *Fundamental and Applied Toxicology*, 15, 320, 1990.

68.   McKee, R. H. and Freeman, J. J., Dermal carcinogenicity studies of petroleum-derived materials, in *Health Risk Assessment - Dermal and Inhalation Exposure and Absorption of Toxicants.* Wang, R. G. M., Knaak, J. B., and Maibach, H. I. (eds.), CRC Press, Inc, 1993.

69.   U.S. Environmental Protection Agency. Risk Assessment Issue Paper: Oral Systemic and Carcinogenic Toxicity for Multiple Fuels, Superfund Health Risk Technical Support Center, Environmental Criteria and Assessment Office, Cincinnati, Ohio, 1992.

70.   U.S. Environmental Protection Agency. On-line Integrated Risk Information System (IRIS), October, 1991.

71.   Agency for Toxic Substances and Disease Registry (ATSDR). 1989. Toxicological Profile for Polycyclic Aromatic Hydrocarbons, Draft, October.

72.   National Library of Medicine (NLM), On-line Hazardous Substances Data Bank (HSDB), October, 1991.

73.   Williams, G. M. and Weisburger, J. H., Chemical carcinogenesis, in *Casarett and Doull's Toxicology - The Basic Science of Poisons.* (Klaassen, C. D., Amdur, M. O., and Doull, J., eds.), Fourth Edition, New York, Macmillan, 1991.

74.   U.S. Environmental Protection Agency, Guidelines for Exposure Assessment; Notice, *Federal Register*, 57(104), 22888, Friday 29, 1992.

75.   Lepow, M. L., Bruckman, L., Gillette, M., Markowitz, S., Robino, R., and Kapish, J., Investigations into sources of lead in the environment of urban children, *Environmental Research*, 10, 415, 1975.

76.   Calabrese, E. J., Barnes, R., Stanek, III, E. J., Pastides, H., Gilbert, C. E., Veneman, P., Wang, X., Lasztity, A., and Kostecki, P. T., How much soil do young children ingest: An epidemiologic study, *Regulatory Toxicology and Pharmacology*, 10, 123, 1989.

77.   Wedeen, R. P., Mallik, D. K., Batuman, V., and Bogden, J. D., Geophagic lead nephropathy: Case report, *Environmental Research*, 17, 409, 1978.

78.   U.S. Environmental Protection Agency, Exposure Factors Handbook, Office of Health and Environmental Assessment, EPA/600/8-89/043, 1990.

79.   Cowherd, C., Muleski, G., Engelhart, P., and Gillete, D., Rapid Assessment of Exposure to Particulate Emissions from Surface Contamination, EPA Office of Health and Environmental Assessment, EPA/600/8-85/002, 1985.

80.   U.S. Environmental Protection Agency, Risk Assessment Guidance for Superfund, Volume I - Human Health Evaluation Manual, (Part B)., Development of Risk-based Preliminary Remediation Goals, EPA/540/11-89/002B, OSWER Directive 9285.7-01B, June 1991.

81.   U.S. Environmental Protection Agency, Development of Advisory Levels for PCBs Cleanup, Office of Health and Environmental Assessment, 1986.

82.   Calabrese, E. J., Kostecki, P. T., and Leonard, D. A., Public health implications of soils contaminated with petroleum products, in *Soils Contaminated by Petroleum: Environmental and Public Heath Effects*, John Wiley and Sons, New York, 191, 1988.

83.   U.S. Environmental Protection Agency, Interim Guidance for Dermal Exposure Assessment, EPA/600/8-91/011A, March 1991.

84.   Ryan, E., Hawkins, E., Magee, B., and Santos, S., Assessing risk from dermal exposure at hazardous waste sites, *Superfund '87 Proceedings of the 8th National Conference*, Washington, D.C., Hazardous Materials Control Research Institute, November 16-18, 1987.

85.   U.S. Environmental Protection Agency, Proposed amendments to the guidelines for the health assessment of suspect developmental toxicants; Request for comments; Notice, *Federal Register*, 54 (42), 9386, 1989.

86.   McKee, R. H., Plutnick, R. T., and Traul, K. A., Assessment of the potential reproductive and subchronic toxicity of EDS coal liquids in Sprague-Dawley rats, *Toxicology*, 46 (3), 267, 1987.

87.   McKee, R. H., Pasternak, S. J., and Traul, K. A., Developmental toxicity of EDS recycle solvent and fuel oil, *Toxicology*, 46 (2), 205, 1987.

88.   Chu, I., Rinehart, W., Hoffman, G., Vileneuve, D. C., Otson, R., and Valli, V. E., Subacute inhalation toxicity of a medium-boiling coal liquefaction product (154-378 degrees C) in the rat [Part III], *Journal of Toxicology and Environmental Health*, 28 (2), 195, 1989.

89.   White, R., Chevron Environmental Health Center, personal communication with Mahini, Xuannga, September 30, 1991.

90.   Clayton, G. D. and Clayton, F. E., *Patty's Industrial Hygiene and Toxicology*, 3rd edition, 1981.

91.   U.S. Environmental Protection Agency, Recommendations for and Documentation of Biological Values for Use in Risk Assessment, Office of Health and Environmental Assessment, Cincinnati, Ohio, EPA/600/6-87/008, 1988.

92.   Agency for Toxic Substances and Disease Registry, Toxicological Profile for Fuel Oils, Draft, May 1993.

93.　Hertzberg, R., Method Evaluation and Development Staff, Environmental Criteria and Assessment Office (ECAO), Cincinnati, Ohio, personal communication with Mahini, Xuannga, November 10, 1992.

94.　U.S. Environmental Protection Agency, Health Effects Assessment Summary Tables - Annual FY-1991, U.S. EPA Office of Solid Waste and Emergency Response, Washington, D.C, 1991.

95.　U.S. Environmental Protection Agency, Health Effects Assessment Summary Tables (HEAST) -Annual FY 1992, March.

96.　U.S. Environmental Protection Agency, Health Effects Assessment Summary Tables (HEAST) -Supplement No. 2 to the March 1992 Annual Update, November, 1992.

97.　U.S. Environmental Protection Agency, Superfund Health Risk Technical Support Center, (513) 569-7300. Toxicity information on naphthalene, March 2, 1993.

98.　U.S. Environmental Protection Agency, National Oil and Hazardous Substances Pollution Contingency Plan; Final Rule, 40 CFR Part 300, 1990.

99.　Seidel, S., U.S. EPA Region IX, personal communication with Martinez, Janet, November, 1990.

100.　U.S. Environmental Protection Agency, Determining Soil Response Action Levels Based on Potential Contaminant Migration to Groundwater, A Compendium of Examples, EPA/540/2-89/057, 1989.

101.　California Regional Water Quality Control Board (CRWQCB), Central Valley Region, The Designated Level Methodology for Waste Classification and Cleanup Level Determination, October 1986, updated June 1989.

102.　Battelle Pacific Northwest Laboratories, Toxicologic Criteria for Defining Hazardous Wastes, Developed for the Minnesota Pollution Control Agency, Roseville, Minnesota, 1976.

103.　U.S. Environmental Protection Agency (EPA), Background Document, Resource Conservation and Recovery Act (RCRA), Subtitle C - Hazardous Waste Management, Section 3001, Identification and Listing of Hazardous Waste, Section 251.24 - EP Toxicity Characteristic, 1980.

104.　California Department of Health Services (CDHS), Leaking Underground Fuel Tank Task Force, Leaking Underground Fuel Tank (LUFT) Field Manual, 1989.

CHAPTER 2

Update on the Derivation of an Oral Reference Dose for Diesel Fuel No. 2

J. E. Ryer-Powder and M. J. Sullivan
ChemRisk Division, McLaren/Hart Environmental Engineering — Irvine, California

INTRODUCTION

Current Reference Dose and Cancer Slope Factor for Diesel Fuel

The United States Environmental Protection Agency (USEPA) developed reference doses (RfDs) in an attempt to aid in risk management decisions. RfDs are defined by the USEPA as "an estimate (with uncertainty spanning perhaps an order of magnitude) of continuous exposure to the human (including sensitive subgroups) that is likely to be without an appreciable risk of deleterious effects during a lifetime." Traditionally, the USEPA has set RfDs by identifying the highest no observed adverse effect level (NOAEL) for the "critical effect" from the scientific literature and dividing by uncertainty and modifying factors which correct the experimental NOAEL to a RfD value which is protective of all members of the human population. In March, 1992, the USEPA developed a provisional oral RfD for diesel fuel of 8 x 10^{-3} mg/kg-day.[1] In addition, Millner, James, and Nye[2] developed a cancer slope factor for Diesel Fuel No. 2 of 4.32 x 10^{-4} (mg/kg-day).[-1] Several concerns have been identified regarding the reference dose and slope factor. USEPA's RfD for diesel fuel is based on studies exposing animals to Marine Diesel Fuel. Marine Diesel Fuel is one specific type of diesel fuel. Diesel Fuel No. 2 has very different chemical characteristics and should therefore be evaluated separately. Regarding the slope factor, the potential for carcinogenicity from dermal exposure to middle distillates has been studied extensively. These studies suggest that chronic irritation is a prerequisite for dermal carcinogenicity. Therefore, development of a cancer slope factor may not be relevant to human exposure to Diesel Fuel No. 2.

The purpose of this chapter is to discuss in further detail the concerns regarding the RfD and cancer slope factor for diesel fuel and propose an updated RfD specific to Diesel Fuel No. 2.

Composition and Classification of Diesel Fuel

Diesel fuels are designed for use as fuels in diesel engines. They are complex hydrocarbon mixtures derived from the refining of crude oil. Diesel fuels are classified as middle distillates, generally coming off the distillation column at temperatures of 150-360 °Celsius (C).

They are more dense and less volatile than gasoline. There are several grades or types of diesel fuels. Those relevant to this report, as described in the US Chemical Substances Inventory under the Toxic Substances Control Act[3] or described by other references are:

Diesel Oil (Chemical Abstract Service (CAS) #68334-30-5) - A complex combination of hydrocarbons produced by the distillation of crude oil. It consists of hydrocarbons having carbon numbers predominantly in the range of C9-C20 and boiling in the range of approximately 163-357°C. It should be noted that this definition encompasses both Diesel Fuel No. 1 and Diesel Fuel No. 2.

Diesel Fuel No. 2 (CAS #68476-34-6) is defined by the US Chemical Substances Inventory as the distillate oil having a minimum viscosity of 32.6 SUS at 38 °C to a maximum 40.1 SUS at 38°C. It is a blend of straight-run and catalytically cracked streams, including straight-run kerosene, straight-run middle distillate, hydrodesulfurized middle distillate and light catalytically and thermally cracked distillates. The boiling range is generally approximately 160-360 °C.[4] Diesel Fuel No. 2 is similar in composition to Fuel Oil No. 2. Some of the polycyclic aromatic hydrocarbons present in Fuel Oil No. 2, and therefore probably present in Diesel Fuel No. 2, include phenanthrene, fluoranthene, pyrene, benz(a)anthracene, chrysene, and benzo(a)pyrene.[4] This grade has a higher specific gravity than Diesel Fuel No. 1, providing more energy per unit volume of fuel and is therefore used in railcars, trucks and river boats.

Diesel Fuel No. 4 (no CAS #) or marine diesel is the most viscous of the diesel fuels and contains higher levels of ash and sulfur.[4] Diesel Fuel No. 4 may contain more than 10% PAHs.[5]

TOXICOLOGY OF DIESEL FUEL

Following a complete review of the epidemiological and toxicological literature with regard to Diesel Fuel No. 2, it was determined that the most sensitive endpoint of toxicity was liver effects. These effects were demonstrated in a study by Chu *et al.*[6] This study is described in detail in the "Derivation of a Reference Dose for Diesel Fuel No. 2" section.

In reviewing the health effects summaries for the diesel fuels and related middle distillates, in many cases, one of the effects is male rat hyaline droplet nephropathy and its sequelae (including renal tumors).[1] This effect is limited in occurrence to male rats, and has been found to be related to the presence of a low-molecular weight protein called alpha 2μ globulin in these animals. Available evidence suggests that humans are not likely to experience these effects.[7] Therefore, hyaline droplet nephropathy and related endpoints (including renal carcinogenicity) were not considered in the development of quantitative oral toxicity values.

Epidemiological studies provided no conclusive evidence for carcinogenicity of diesel fuels to humans. Skin painting assays in animals have reported positive results for some of these fuels, but this response is suggested to be due to epigenetic processes related to skin irritation,[8] and therefore, not necessarily relevant to exposure by other routes.

USEPA'S REFERENCE DOSE FOR DIESEL FUEL

USEPA's RfD was derived as follows.[1] A subchronic inhalation study by MacEwen and Vernot[9] was conducted in which continuous 90-day inhalation exposure to 50 or 300 mg/m3 of marine diesel fuel derived from petroleum or shale resulted in fatty change in the livers of female mice. A LOAEL of 50 mg/m^3 was identified based on this result. Assuming equal absorption by the inhalation and oral routes and using standard reference values[10] for female C57BL/6 mouse body weight (0.0246 kg) and inhalation rate (0.040 m^3/day), the equivalent oral dose was calculated as follows:

50 mg/m^3 x 0.040 m^3/day x 1/0.025 kg = 81 mg/kg-day

Applying an uncertainty factor of 10,000 to reflect (1) variation within species (2) variation between species (3) extrapolation to chronic duration, and (4) use of a LOAEL instead of a NOAEL, a RfD of 0.008 mg/kg-day was derived.

In reviewing the physical and chemical characteristics associated with marine diesel fuel and Diesel Fuel No. 2, it is apparent that the two fuels are very different. Table 1 describes these differences in detail.

In examining this table with regard to parameters that may effect the toxicity of the diesel fuels, it is evident that Diesel Fuel No. 4 (marine diesel fuel) contains a greater percentage of aromatics and 3-7 member rings. In a presentation on the toxicity associated with chemical component classes of refinery streams by Feuston et al.,[11] the authors correlated chemical components with endpoints of toxicity for several refinery

Table 2-1   Characteristics of Diesel No. 2 vs Diesel No. 4[1]

Characteristic	Diesel No. 2	Diesel No. 4
Hydrocarbons[a]	68-75 % paraffins 18-25 % aromatics	12.7% paraffins 87 % aromatics (NTP)
Concentration in Fuel (mg/g)[b] benzene	.01	0.02
toluene	4.7	7
ethylbenzene	0.26	0.25
m- and p-xylenes	1.0	0.20
styrene	<0.06	0.66
o-xylene	0.32	<0.02
n-propyl benzene	0.15	0.24
1,3,5 trimethyl benzene	0.87	0.12
		0.43
Benzo[a]Pyrene (ug/g)[b]	0.05 - 0.84	0.03 - 0.09
3 - 7 Membered Rings[a]	< 5%	> 10%

[1] The above table was compiled from (a) IARC Monograph 45 and (b) Griest et al., (1986) Comparative Chemical Characterization of Shale Oil and Petroleum Derived Diesel Fuels. DE86003310, Oak Ridge National Laboratory.

streams. Streams containing 3-5 membered rings or 4-5 membered rings were associated with an increased liver weight and increased liver pathology. In reviewing the literature, it was apparent that exposure to either Diesel No. 4 or Diesel No. 2 causes changes in the liver.[6,9,12,13,14] In a case study,[15] ingestion of 28,000 mg/kg resulted in blood chemistry changes which the authors attributed to kidney and liver toxicity. A very mild histological change occurred in the liver of rats exposed to 100 mg/kg of an aerosol of Diesel Fuel No. 2 for 6 hr/day, 5 days/week, for 4 weeks.[6] A dose-dependent increase in liver necrosis occurred with dermal exposure of rabbits to 4 ml/kg or 8 ml/kg, 5 days per week for 2 weeks.[13] Yagaminas *et al.*,[12] demonstrated that dermal application to rats of 1000 mg/kg, 5 days/week for 2 weeks of diesel fuel results in a decreased liver weight, decreased cholesterol content, and decreased serum protein content. MacEwen and Vernot[9] demonstrated a fatty change and inflammation in the liver of female mice following continuous exposure for 90 days to 50 or 300 mg/m^3 of marine diesel fuel. In a chronic dermal carcinogenesis study conducted by the American Petroleum Institute,[13] 0.05ml Diesel Fuel No. 2 applied to the backs of mice 3 times per week for 62 weeks resulted in pathological disorders of the liver and other organs.

SUGGESTED CANCER SLOPE FACTOR FOR DIESEL FUEL NO. 2

Recently, Millner *et al.*,[2] developed a soil cleanup guideline for Diesel Fuel No. 2 based on the dermal carcinogenic potential of diesel fuel and its constituents. Several studies suggest that middle distillates are tumor promoting substances as opposed to tumor initiating substances.[8,16-19] Light and intermediate middle distillates do not appear to have mutagenic potential.[20] Chronic skin irritation is strongly suspected as being a necessary prerequisite for tumor promotion by many substances.[21-24] It is not necessary for the chronic skin irritation to be severe in order to promote tumor development. In fact, severe irritation seems to decrease the tumor response by being fatal to initiated cell populations.[25] It has been demonstrated that suppression of chronic irritation in mouse skin can also suppress the tumor response.[26-29]

Chronic exposure to middle distillates has been shown to produce skin irritation in rodents.[20,30-31] Upon prolonged application, Diesel Fuel No. 2 as well as many other petroleum hydrocarbons, may produce a sufficient degree of chronic irritation in initiated mouse skin sufficient to promote tumor development. However, mice and humans may respond differently to prolonged application of different doses of middle distillates since the skin of these two species is different. No evidence exists to suggest that petroleum middle distillates are tumor initiators. In setting a reference dose for a chemical, it is necessary to consider the levels of expected exposure. In the case of Diesel Fuel No. 2, it is not likely that human exposure would be on the order of that required to produce a chronic irritation. No data was available for oral or inhalation cancer bioassays. Therefore, based on the available data and the IARC[4] evaluation for carcinogenicity of light diesel fuels, this chemical was evaluated as a non-carcinogen.

DERIVATION OF A REFERENCE DOSE FOR DIESEL FUEL NO. 2

In deriving a RfD for Diesel Fuel No. 2, a similar rationale to USEPA's[1] for choosing an end-point of toxicity, i.e., liver effects, was used. However, EPA chose a study using marine diesel fuel. As previously explained, this fuel is significantly different from Diesel Fuel No. 2; therefore, two studies exposing animals to Diesel Fuel No. 2 were chosen in order to derive a Reference Dose (RfD) specifically for Diesel Fuel No. 2.

The first of these studies is a two-year dermal carcinogenicity study. It is difficult to derive a RfD from dermal studies because data regarding the bioavailability, i.e., the percentage of a chemical that is absorbed into the body, of the compound is usually not available. However, in the absence of chemical data, an approach similar to that used by the State of Michigan could be adopted.[32] In that state, dermal bioavailability is assumed to be 1% for semivolatiles and metals. It should be noted that a dermal bioavailability of 1% is extremely conservative because it assumes that the no observable adverse effect level (NOAEL) or lowest observed adverse effect level (LOAEL) occurred from 1% of the dose that was dermally applied.

In their investigation, Clark *et al.*,[33] investigated the dermal carcinogenic potential of petroleum and hydrotreated shale oils in C3H/HeN mice. Ten test materials (including petroleum and shale diesel) were applied 3 times per week for up to 105 weeks to the shaved skin of mice. Mineral oil and benzo(a)pyrene (0.15% weight volume) were used as negative and positive controls, respectively. The test groups consisted of a total of 50 subjects with equal numbers of males and females. Approximately 25 mg of test material was applied to the shaved dorsal thoracic region of each animal. Inflammatory lesions at the site of application, septicemia, and dehydration resulted in increased mortality in a number of the test groups. Histopathological examinations were conducted on skin, internal organs, and gross lesions upon death of the animals. Chronic toxicity of the test materials was limited to inflammatory and degenerative skin changes.

This 2-year study was used to select a NOAEL for Diesel Fuel No. 2. A RfD can be derived as follows:

1. To account for 1% absorption:
   25 mg x 0.01 = 0.25 mg

3. To adjust for a 3 day/week exposure and body weight:
   0.25 mg/day x 3 days/7 days x 1/0.025 kg = 4.28 mg/kg-day

4. Divide by an uncertainty factor of 100 (variation within species, variation between species)
   4.28/100 = 0.043 mg/kg-day

Therefore, the RfD for Diesel Fuel No. 2 is 0.04 mg/kg-day.

An inhalation study using Diesel Fuel No. 2 can be used to support this RfD. In a study conducted by Chu *et al.*,[6] rats were exposed to an aerosol of a medium boiling coal liquefaction product (CLP) at 25 mg/m^3 or 100 mg/kg for 6 hours per day, 5 days per week for 4 weeks. A positive control received Diesel Fuel No. 2 aerosols at 100 mg/m.^{3} Very mild histological changes occurred in the liver and thyroid of rats treated with both the CLP and diesel fuel. According to the authors, the changes in the liver were adaptive in nature and the thyroid changes were confined to reduced follicular size and angularity, increased epithelial height, and cytoplasmic vacuolation. No treatment-related changes were found in the other tissues examined. No other biochemical or hematological effects were observed in animals exposed to diesel fuel.

The results of this study suggest an animal NOAEL of 100 mg/m.^{3} An animal dose, in mg/kg-day was derived assuming a rat respiratory rate of 0.26 m^3/day of air and a body weight of 0.2 kg. A duration of 6 hours/day, 5 days/week was used to extrapolate a continuous exposure from an intermittent exposure:

$$100 \text{ mg/m}^3 \text{ x } 0.26 \text{ m}^3/\text{day x } 1/0.2 \text{ kg x } 6/24 \text{ x } 5/7 = 23.1 \text{ mg/kg-day}$$

Applying an uncertainty factor of 1000: a factor of 10 for converting an animal NOAEL to a human NOAEL, a factor of 10 to protect the more sensitive individuals of a human population, and a factor of 10 to account for the extrapolation of the exposure duration from subchronic to chronic results in a human RfD of 0.02 mg/kg-day. An underlying assumption is equal bioavailability through inhalation and oral exposure routes.

This RfD is on the same order as the 0.04 value obtained from the chronic dermal carcinogenesis study. Because the dermal study was of a longer duration, the RfD value of 0.04 is more appropriate to use for human risk assessment purposes.

CONCLUSIONS

Based on a review of the literature, an updated RfD specific to Diesel Fuel No. 2 has been proposed. A provisional RfD set by USEPA for diesel fuel is actually based on studies with marine diesel fuel. Marine Diesel Fuel is different from Diesel Fuel No. 2 in several chemical and physical characteristics. A cancer slope factor has also been suggested for Diesel Fuel No. 2. The cancer slope factor is based on studies demonstrating skin tumors following prolonged and repeated application of middle distillates to mouse skin. As chronic skin irritation appears to be a prerequisite for skin tumors, it was determined that these studies are not relevant to the human exposure situation. The current RfD of 0.04 mg/kg-day is based on toxicological studies specific to Diesel Fuel No. 2 and is therefore more appropriate to used for human risk assessment purposes.

REFERENCES

1.   USEPA. Oral Reference Doses and Oral Slope Factors for JP-4, JP-5, Diesel fuel, and Gasoline. Memorandum from Joan S. Dollarhide to Carol Sweeney. Offic of Research and Development - Environmental Criteria and Assessment Office. Cincinnati, Ohio 45268. March 24, 1992.

2.   Millner, G.C., James, R.C., and Nye. A.C. Human Health-Based Soil Cleanup Guidelines for Diesel Fuel Number 2. Journal of Soil Contamination 1992:1(2):103.

3.   Health and Safety Regulation Committee Task Force on Toxic Substances Control. *Petroleum Process Stream Terms Included in the Chemical Substances Inventory Under the Toxic Substances Control Act (TSCA).* American Petroleum Institute. 1983.

4.   IARC. Monographs on the Evaluation of Carcinogenic Risks to Humans. Occupational Exposures in Petroleum Refining; Crude Oil and Major Petroleum Fuels. Volume 45. World Health Organization. 1989.

5.   CONCAWE. 1985. Health Aspects of Petroleum Fuels. Potential Hazards and Precautions for Individual Classes of Fuels (Report Number 85/51). As cited in IARC, 1989.

6.   Chu I, Rinehart W, Hoffman G, Villeneuve DC, others. Subacute inhalation toxicity of a medium-boiling coal liquefaction product (154-378 degrees C) in the rat (Part III) J *Toxicol. Environ. Health* 1989; Vol 28,Iss 2:195.

7.   U.S. EPA. Alpha 2u-globulin: Association with chemically induced renal toxicity and neoplasia in the male rat. Prepared for the Risk Assessment Forum, U.S. Environmental Protection Agency, Washington, DC. 1991. EPA/625/3-91/019F

8.   McKee RH, Plutnick RT, Przygoda RT. The carcinogenic initiating and promoting properties of a lightly refined paraffinic oil. *Fund. Appl. Toxicol* 1989;12(4):748.

9.   MacEwen JD, Vernot EH. Toxic Hazards Research Unit Annual Technical Report. NTIS. AD-A161-5582. 1985:202.

10. U.S. EPA. Recommendations for and documentation of biological values for use in risk assessment. Prepared by Environmental Criteria and Assessment Office, U.S. Environmental Protection Agency, Cincinnati, OH. Prepared for Office of Solid Waste and Emergency Response, Washington, DC. 1987. EPA 600/6/87-008

11. Feuston MH, Mackerer CR, Mehlman MA. Toxicity Associated with Chemical Component Classes of Refinery Streams. Mobil Oil Corporation, Princeton, N.J. Society of Toxicology, 1990.

12. Yagminas A, De Vries PA, Villeneuve DC. Systemic Toxicity of Coal Liquefaction Products: Results of a 14-day Dermal Exposure. *Bulletin of Environmental Contamination and Toxicology.* 1988:40(3):433.

13. American Petroleum Institute. Project Numbers PS-33 & PS-39, Report Abstract Numbers - 27-32817 & 28-31346 1980a.

14. American Petroleum Institute. Project Number PS-36, Report Abstract Number -30-31646. 1983.

15. Parker GA, Bogo V, Young RW. Acute Toxicity of Petroleum- and Shale Derived Distillate Fuel, Marine; Light Microscopic, Hematologic, and Serum Chemistry Studies. *Fund. and Appl. Tox.* Jul., 1986:7(1):101.

16. American Petroleum Institute (API). Short-term dermal tumorigenesis study of selected petroleum hydrocarbons in male CD-1 mice, initiation and promotion phases. API Health and Environmental Sciences Department report 36-32643. 1988.

17. Gehart, J.M., Hatoum, N.S., Halder, C.A., Warne, T.M., and Schmitt, S.L.. Tumor initiation and promotion effects of petroleum streams in mouse skin. *Fundam. Appl. Toxicol.* 1988:11,76.

18. U.S. Environmental Protection Agency, Office of Toxic Substances. Preliminary Evaluations of Initial TSCA Section 8 (e) Substantial Risk Notices, January 1987 - December 1988: Report 8EHQ-1288-0773, 420. 1988.

19. Marino, J.J., Finkbone, H.N., Strother, D.E., Mast, R.W. and Furedi. Tumor initiation and promotion activity of straight run and fractionated kerosenes. Prenented at the 1990 Society of Toxicology meetings, Moami, Fl., Abstract #776. 1990.

20. American Petroleum Instutute. Results of toxicological studies performed under contract for the API, medical and biological science department. 1989.

21. Frei, J.V. and Stephens, P. The correlation of promotion of tumor growth and induction of hyperplasia in epidermal two-stage carcinogenesis. *Brit. J. Cancer*, 22:83, 1968.

22. Argyris, T.S. Tumor promotion by abrasion induced epidermal hyperplasia in the skin of mice. *J. Invest. Dermatol.* 75:360, 1980.

23. Argyris, T.S., and Slaga, T.J. Promotion of carcinomas by repeated abrasion in initiated skin of mice. *Cancer Res.* 41:5193, 1981.

24. Hergenhahn, M., Rurstenberger, G., Opferkuch, J.J., Adolf, W., Mack, H. and Hecker, E. Biological assays for irritant Tumor-initiating and promoting activities. *J. Canc. Res.* 104:31, 1982.

25. Hennings, H. and Boutwell, R.K. Studies on the mechanism of skin tumor promotion. *Cancer Res.* 30:312, 1970.

26. Skisak, C.M. The role of chronic irritation and inflammation in tumor promotion in CD-1 mice by petroleum middle distillates. *Toxicol. Appl. Pharmacol.,* `990.

27. Belman, S. and Troll, W. The inhibition of croton oil-promoted mouse skin tumorigenesis by steroid hormones. Cancer Res. 32:450, 1972.

28. Ghadially, F.N. and Green, H.N. The effect of cortisone on chemical carcinogenesis in the mouse skin. *Brit. J. Cancer.* 8:291, 1954.

29. Slaga, T.J. and Scribner, J.D. Inhibition of tumor initiation and promotion by anti-inflammatory agents. *J. Natl. Cancer Inst.,* 51:1723, 1973.

30. Hoekstra, W.G. and Phillips, P.H. Effects of topically applied mineral oil fractions on the skin of guinea-pigs. *J. Invest. Dermatol.* 40(2):79, 1963.

31. World Health Organization (WHO) Selected Petroleum Products, Environmental Health Criteria 20, Geneva, pp. 11-114, 1982.

32. Michigan Council on Environmental Quality (Michigan CEQ). State of Michigan Draft Risk Assessment Guidelines. Committee on Risk Assessment, Lansing, MI. May 17, 1990.

33. Clark, CR, Walter, Mk, Ferguson, PW, Katchen, M. Comparative dermal carcinogenesis of shale and petroleum-derived distillates. *Toxicol. Ind. Health.* 1988;4:11.

CHAPTER 3

Viability Of Hydraulic Sweeping Of Bunker C And Diesel Fuel From Alluvial Sediments

Paul D. Kuhlmeier,Ph.D.,P.E.
Southern Pacific Transportation Company — Boise, Idaho

Michael J. Grant
Southern Pacific Transportation Company — San Francisco, California

INTRODUCTION

Dating back to the turn of the 20th century, the Southern Pacific Transportation Company's (SPTCo) Dunsmuir Yard represents one of the oldest active industrial facilities in northern California. The Yard is located approximately 45 miles north of Redding, California along the Sacramento River. A general schematic of the site is shown on Figure 3-1. Throughout its history, the facility has played many roles including repair and maintenance of freight and passenger trains, passenger transfer, fueling, and switching. In order to support these activities, several fuel and oil tanks have been built and removed, both above grade and below. Primarily as a result of the use of these tanks and associated surface fueling, petroleum has been released into the alluvial sediments beneath the site. Railroad petroleum problems are often unique in the sense that western railroads used heavy Bunker C fuel oils for decades up to the 1950's.

Upon weathering, Bunker C oil is a highly viscous and largely nontoxic type of petroleum. Nevertheless, cleanup criteria are often derived for Bunker C oil in the same manner as gasoline. The Dunsmuir site contains a significant mass of Bunker C oil trapped in porous sands and gravel. The distribution of resident Bunker C as it exists today is shown on Figure 3-2. It was postulated that such porous conditions (hydraulic conductivity > 10^{-1} cm/s) would be conducive to potential hydraulic extraction. Three discrete modeling efforts were accomplished to support engineering design. Ground water flow and containment options were simulated with the U.S. Geological Survey MODFLOW model. Hydrocarbon mobility and extractability was assessed by empirical modeling supported by bench-scale feasibility testing with surfactants. The studies performed to evaluate hydraulic removal are the focus of this chapter.

In order to fully assess the extent of petroleum beneath the site, identify potentially critical source areas, and establish engineering design constraints, a comprehensive field effort was completed. A preliminary understanding of the geologic environment was extended to allow for accurate assessment of plausible engineering alternatives and

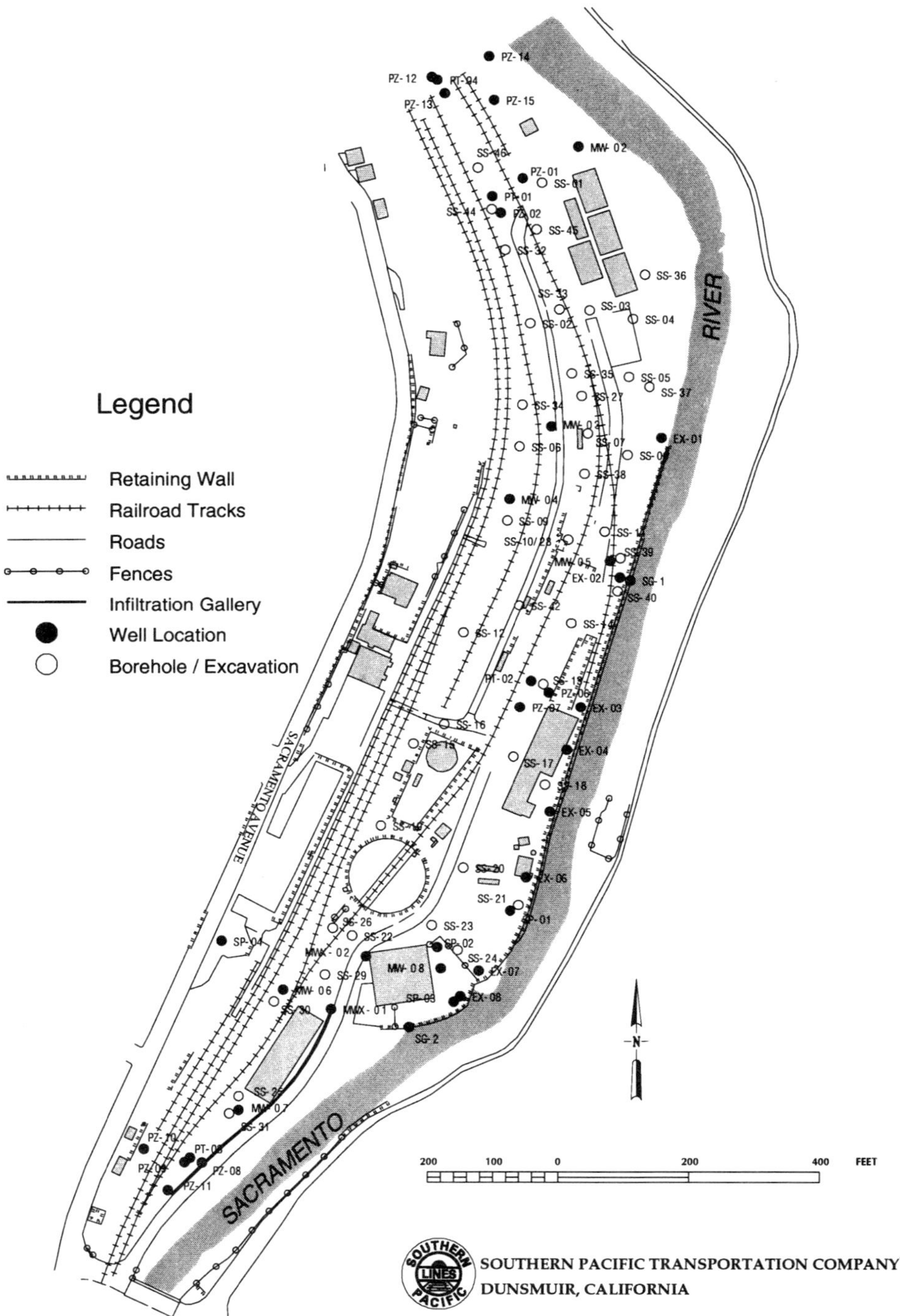

Figure 3-1   Dunsmuir Yard, Site Map

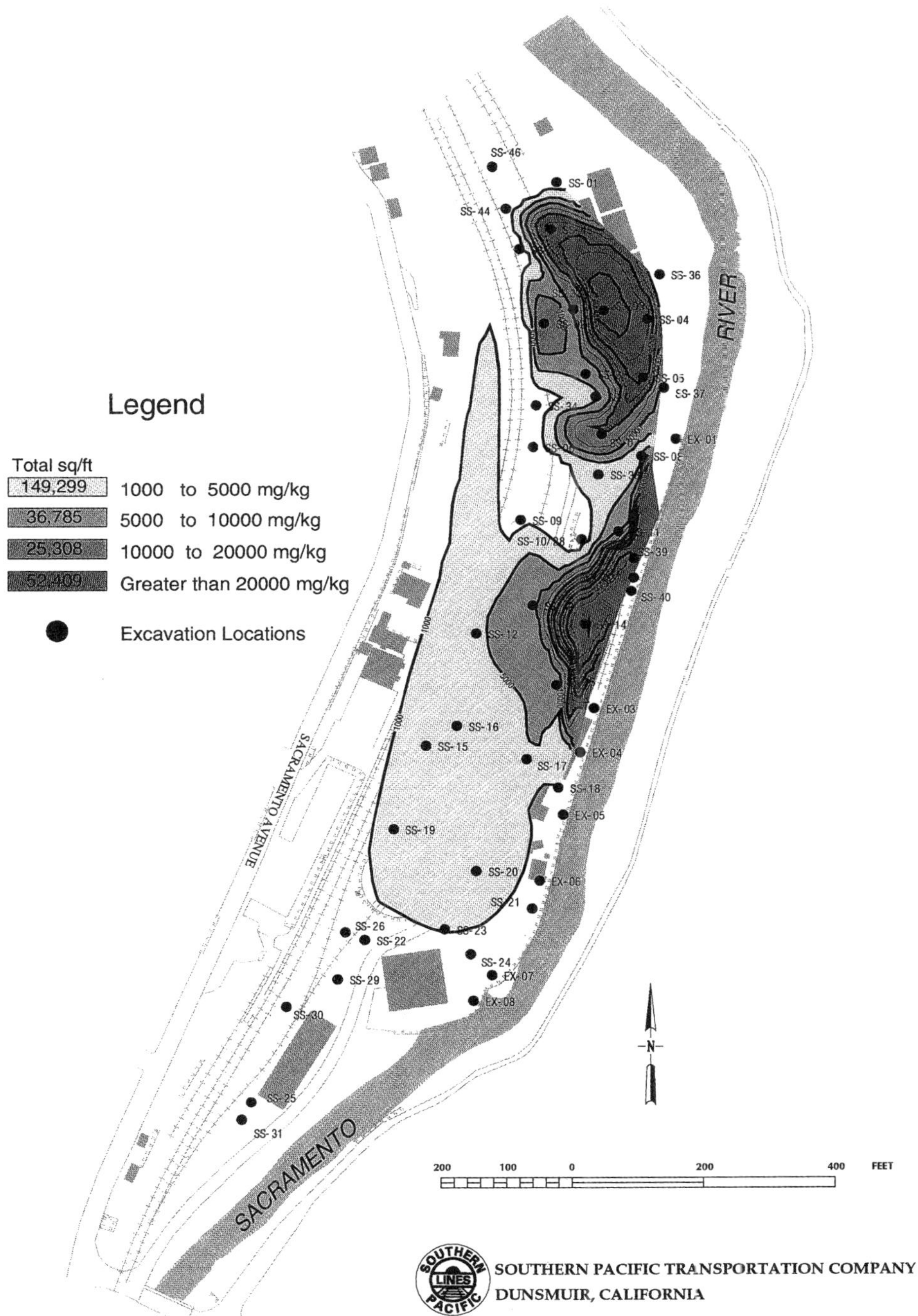

Figure 3-2　Dunsmuir Yard, 1991 and 1992, TPH in Alluvial Sediments

to provide input for hydraulic sweeping analysis. In total, 16 excavations were made to the groundwater table for physical and chemical cataloging. This series of test pit work brought the total number of such excavations to more than 40 across the site. Hydrocarbons were speciated in the laboratory providing the basis for determining the extractability and mobility of pollutants in the subsurface. Over two-thirds of the resident hydrocarbon mass is in the form of very long chain n-alkanes found in the paraffin and wax family of chemicals, here remnants of weathered Bunker C oil. Two distinctive fingerprints of concentrated weathered heavy oil were identified in the northeast and central portions of the site (Figure 3-2). These critical areas became the design basis for the engineered system.

Field and laboratory tests were conducted to elucidate seven key physical properties of site sediments. These tests included grain size, moisture content, bulk density, dry density, specific gravity, porosity, and total organic carbon content. Results demonstrated that a significant amount (>1 percent) of natural organics are present in sediments, which tend to retard or adsorb diffuse petroleum products upon contact.

Groundwater characterization consisted of monitoring and pumping well installation, long-term aquifer tests, dye tracer tests, and ground water quality sampling. In all, ten wells were installed and two 72-hour aquifer pumping tests were performed. The porous alluvial gravel and boulders produced an unusually low volume of water, sustaining less than 10 gpm with only minor lateral drawdown. This observation meant that if groundwater was to be captured, classical well systems would not be appropriate, and thus a line sink would be considered. Dye tracer tests proved that subsurface water moves rapidly towards the river at over 100 ft/day, but that the shallow aquifer thickness does not contribute large volumes of water under base flow conditions. Water quality sampling from 11 wells indicated an average TPH concentration of 3.6 mg.

HEAVY OIL FATE AND TRANSPORT

Theory

Migration and extraction of NAPLs, particularly very long chain aliphatics and aromatics, is a complex process. Transport through the vadose zone occurs as a continuous multi-phase flow function under the influence of capillary, viscous, and gravity forces. Once the source of hydrocarbon has stopped and the bulk of hydrocarbon displaced, some of it is trapped in the porous media because of capillary forces. Hydrocarbon migration stops as this lower, residual saturation is reached. It is this mass of hydrocarbon with which extraction and cleanup becomes a serious challenge. This challenge to remove the resident mass is further exacerbated when one is dealing with highly weathered heavy fuel oils, deposited decades in the past.

Since the organics of interest at the Dunsmuir Yard are lighter than water, the predominate mass has spread laterally on the water table and has subsequently smeared throughout the interval of seasonal water table fluctuation. In the case of this site, the smear zone amounts to two to five feet in length. Interfacial forces acting between the water phase or air phase and the NAPL have caused residual "blobs" of the organic to be retained in the smear zone. Migration of historical spills have probably ceased at the site as most of the bulk organic phase has become trapped in discontinuous blobs based on 32 sample observations from test pits and some 27 extracted soil sample analyses.

Documentation of the existence of NAPLs has become a frequent topic of research (Atwater, 1984; Cohen et al., 1987; Feenstra and Coburn, 1986; Schwille, 1988), and the need for an enhancement of understanding of the fate and transport processes associated with petroleum hydrocarbons is a priority with the U.S. Environmental Protection Agency. Several theoretical models have been developed to describe multi -phase flow of petroleum hydrocarbons in the subsurface with the exchange of organic mass between fluid phases (Abriola and Pinder, 1985; Corapcioglu and Baehr, 1987; Baehr, 1987; and Dorgarten and Tsang, 1990). None of these however, have been rigorously tested under field conditions.

There are several limitations to the models which have been applied in hypothetical and practical cases. Perhaps the most prominent is assumption of local equilibrium. The local equilibrium assumption implies that if the concentration of contaminant in one phase is known, its concentration in other phases at the same spatial location can be described by equilibrium partitioning relationships (Abriola and Pinder, 1985). It is also noted that spatial location is defined on the macroscopic scale and does not consider immobile phase transfer limitations. There is some evidence that suggests that the local equilibrium assumption does provide a plausible if not overly conservative upper bound to NAPL phase exchange, and thus extractability (Fried et al., 1979; Pfannkuch, 1987; Anderson et al., 1982; and Hunt et al., 1988).

Data collected to date at the Dunsmuir Yard appears to demonstrate that a non-equilibrium description is required for characterization of fluid-fluid mass transfer for long chain petroleum hydrocarbons. Groundwater concentrations beneath the site are more than one order of magnitude below equilibrium levels that would be expected given the application of the previously referenced models. This condition was not unexpected however. Even short chain hydrocarbons such as benzene and toluene downstream from a large petroleum blob similar to the Dunsmuir condition, were found by Geller and Hunt (1989) to be significantly below thermodynamic equilibrium conditions, observations which appear to refute the local equilibrium assumption. Irrespective of the theoretical implications, it is important to describe the dynamics of this complex partitioning in manner that can be articulated to the public and applied in practice. An attempt is made here to present and utilize basic engineering principles that can functionally describe the constraints of this problem.

Rate Limited Sorption

Both site-specific data and data reported by others demonstrate the importance of the sorption limiting phenomena when attempting to extract or even predict the movement of heavy oil hydrocarbons. It has been theoretically and experimentally shown that rate-limited sorption/desorption can have a profound effect upon the transport of sorbing organic chemicals (van Genuchten and Wierenga, 1976; Goltz and Roberts, 1987; Brusseau and Rao, 1989). The advection/dispersion equation, which has been traditionally used to model contaminant transport, uses a retardation factor to account for sorption, thereby implicitly assuming local equilibrium between contaminant in the sorbed and aqueous phases (Bear, 1979). By making this assumption, the possibly large effects of rate-limited sorption/desorption are not considered. For ease of mathematical manipulation, linear sorption expressions are most often used. This assumption which implies an infinite amount of solute/sorbant sites exist leads to large errors in mass removal estimates. The application of Freundlich, BET, or Langmuir isotherms appear to be more appropriate for heavy oil sorption response.

If we assume that extraction methods of some type at the site will be accomplished through wells or trenches, we can then evaluate the effects of rate-limited sorption through the application of radial flow geometry. Various workers have presented models describing radial transport of organics into and out of an aquifer from injection/extraction wells. Most of these models also have been formulated assuming advective/dispersive transport of a nonsorbing solute (Orgata, 1958; Moench and Ogata, 1981). Unfortunately, the referenced models are applicable for describing sorbing solute transport if sorption is governed by a linear, reversible, equilibrium isotherm.

Recently developed simulation techniques coupled with a growing body of field observation data have led to rate limitations being fit to first-order mechanism (Valocchi, 1986) in coarse grained soils similar to those encountered at the Dunsmuir Yard. This work suggests the Freundlich type sorption isotherm should be applied. Based on the site data collected to date, and the body of literature, it is evident that the extraction function lies between a linear form (overestimating movement) and a logarithmic decay (under estimating movement at residual saturation). It is noted that applying the equations of Valocchi (1986) and extending the sorption term to a second order rate function, as the oil content declines below residual saturation, even a logarithmic decay function appears to significantly overestimate desorption, ergo, extractability of the weathered Bunker C oil at the Dunsmuir Yard, specifically.

Field Observations

Residual saturation is typically expressed as the fraction of the void space occupied by NAPL. One may best describe residual saturation as a function of site geotechnical properties and the physical properties of the solute. It is instructive to first review the plausible ranges of residual petroleum saturation observed by others as a basis for comparison to specific site conditions. McKee et al. (1972) measured residual gasoline saturation in the variably saturated zone at 7.7 percent after the water table level dropped and the column drained for 21 days. Dietz (1980) reported residual saturations in coarse porous media of 10 percent for light oil and 20 percent for heavy oil. Thornton (1980) measured gasoline residual saturations of 23 to 29 percent after flushing with water. Schwille (1984) found residual oil saturations in the range of 0.71 to 1.25 percent in highly permeable media and 7.5 to 12.5 percent in low permeability soils. Gasoline residual saturation in unsaturated sand has been observed to range from 12 percent in medium and coarse-sized initially water-wet sand to 60 percent in initially dry, fine sand (Hoag and Marley, 1986).

An effort was made to evaluate the site data with respect to probable residual saturation. Observations from on-site test pits reveal that heavy petroleum hydrocarbons are found at concentrations between 0.5 and 10 percent. Oil taken from a site wastewater plant well which collects groundwater beneath the oil zones was passed through four discrete 2 ft^3 samples of contaminant free site sediments. Clean water was passed through the samples under gravity drainage conditions. In all, 30 pore volumes of water was used at a rate of two pore volumes per day. After four weeks of allowing the material to sit undisturbed, sediment subsamples were collected and analytical results confirmed oil concentrations between 1.5 and 8 percent remaining in the samples. The selection of four weeks was based solely on the time constraints for project completion, however this period was considered reasonable for estimating an equilibrium condition. This elementary study further suggests the oil is at or near residual saturation beneath the site as it relates to the heavy oil component.

Mass Transfer Relationships

Once we gained a better understanding of the relative mass of oil in the ground and its relationship to residual saturation we then needed to assess how the oil may migrate. Bulk advection capability has been depleted, thus mass transfer rates for exchange of weathered oil from pure organic phases to an aqueous phase are then considered to be limited primarily by diffusion of the constituent through a boundary layer from the interface of the bulk aqueous phase (deZabala and Radke, 1986; Pfannkuch, 1984). For this work it was assumed that a single-resistance, linear-driving-force model adequately describes the rate of such interphase mass transfer; thus the flux in a direct ion normal to the interface between the two phases is:

$$F_i^o = -k_{I_i}(C_i - C_{I_i})$$

where k_{I_i} is the mass transfer coefficient for solute i across the boundary layer and C_i and C_{I_i} are the mass concentrations of i in the bulk aqueous solution and at the interface respectively. The mass transfer coefficient (k_{I_i}) is directly proportional to the free liquid diffusivity of the solute and inversely proportional to the thickness of the boundary layer. Under ordinary circumstances it is not physically possible to measure the solute concentration at the interface, and it is therefore convenient to define the mass transfer coefficient on the basis of the equilibrium concentration of species i in the aqueous phase (C_{si}). For a pure NAPL, the equilibrium concentration of a species is the solubility of NAPL in water. Then,

$$F_i^o = -k_{f_i}(C_i - C_{S_i})$$

where k_{f_i} is an effective mass transfer coefficient.

Several correlation equations for mass transfer coefficients for transport through liquid boundary layers at low velocities are compiled in Table 3-1. These empirical relationships are typically based on a correlation for the Sherwood number Sh, a dimensionless parameter relating interphase mass transport resistances to molecular mass transport resistances (Welty et al., 1969):

$$Sh = K_{f_i} l_c / D_{l_i} = a + bRe^m Sc^n$$

Here Re and Sc are the Reynolds and Schmidt numbers, respectively; and a, b, n and m are empirical constants.   Many of these relationships were derived from measurements of dissolution of organic spheres by a homogeneous flow field in a packed bed.   Some have been developed for diffusion-limited dissolution of fluid spheres. These correlations are therefore not directly characteristic of the dissolution of a system of NAPL blobs in porous media.   Correlations developed for single fluid spheres (Freidlander, 1957; Bowman et al., 1961), do not account for the tortuous flow around blobs caused by the presence of a solid matrix. Tortosity would be expected to reduce the boundary layer thickness around blobs and thus increase the value of the mass transfer coefficient (Pfannkuch,1984).   These relationships are particularly important at the site due to high turbulence observed in aquifer pump wells which indicate laminar flow assumptions (very low Reynolds number) may not be valid. Thus the inclusion of relationships that are Reynolds number sensitive are appropriate for use at this site.

Table 3-1   Mass Transfer Coefficient Correlations for Packed Beds and Flow Around Single Spheres

References	Equation[1]	Characteristic Lenght, 1_L	Valid Conditions
Wilson and Geankoplis (1966)[2]	$Sh=1.09(ES_W)^{-1}Pe)^{0.33}$	d_b	$0.0016<Re<55$ $0.35<ES_W<0.75$ $950<Sc<70600$
Dwwievedi and Updhyay (1977)[3]	$Sh=1.1068(ES_W)^{-1} Re^{0.28}Sc^{0.33}$ $Sh=Sc^{0.33}(ES_W)^{-1} (0.765 Re^{0.18}+0.365Re^{0.614})$	d_b	$Re<10$
Kumar et al. (1977)[4]	$Sh=1.110(ES_W)^{-1}Re^{0.2814}Sc^{0.33}$	d_b	$0.016<Re<10$ $0.26>ES_W<0.632$ $123<Sc<70600$
Williamson et al. (1963)[2]	$Sh=250Pe^{0.33}$ $Sh=2.40(ES_W)0.67Re^{0.33}Sc^{0.42}$	d_b d_b	$0.036 < Re < 55$ $0.08 < Re (ES_W)^{-1}>125$
Pfannuch (1984)[5]	$Sh=0.55+0.25Pe^{1.5}$	d_g	$0.5<Pe<100$
Friedlander (1957)[6]	$Sh=0.89Pe^{0.33}(ES_W)^{-0.33}$	d_b	$Re<5,Pe(ES_W)^{-1}>1000$
Bowman et al.(1961)[6]	$Sh=2$ $Sh=0.978Pe^{0.33}(ES_W)^{-0.33}$	d_b d_b	$Pe<1$ $Pe>10$

[1] Sh, Sherwood number, $(Sh=kf_i l_c/D_{l_i})$; Re, Reynolds number, $(Re=1_c qp_w/\mu_w)$; Sc, Schmidt number , $(Sc = \mu_w /(D_{l_i} p_w))$; Pe, Peclet number, $(Pe=ReSc)$; d_b, Longest blob dimension (L); d_g, Sand grain diameter (L).
[2] Solid spheres in packed bed.
[3] Solid spheres, cylinders and flakes in packed and fluidized beds.
[4] Solid cylinders in packed beds.
[5] Oil dissolution in porous media.
[6] Flow around single spheres.

Several investigators who have studied mass transfer in porous media have assumed that blobs are distributed as discrete spherical droplets (deZabala and Radke, 1986; Hunt *et al.*, 1988; Pfannkuch, 1984). Hunt *et al.* (1988) used spherical blobs with diameters of 0.001 to 10 m and found that nonequilibrium conditions would be observed over many of the large blob sizes considered. In the work by deZabala and Radki (1986), blobs were modeled as very small (1 x 10^{-5} m) spheres. They concluded that the local equilibrium assumption would be valid in all cases considered. Qualitative analysis of 16 site soil samples was performed to assess if blobs could be identified and if so, whether they could be measured. Care was taken to analyze samples from high, medium, and low concentrations of Bunker C impacted soil and diesel impacted soil (sediments). Discrete blobs were apparent in both Bunker C oil and diesel oil affected soils. The range in blob size was less than 0.01m to an estimated 0.5m for Bunker C affected soils and less than 0.01m to 0.04m for diesel affected soils. Diesel blobs were distinguishably smaller than commensurate Bunker C soil blobs given similar TPH concentrations. While the analyses were qualitative in nature, they do provide insight as to the distribution of long chain hydrocarbons (C_{20} and above) compared to shorter chain hydrocarbons. Insight is also gained in determining the real range of blob sizes that should be applied in modeling. Discrete blob sizes greater than 0.5m were not observed even in the highly porous rivertine soils/sediments at the Dunsmuir Yard. Rather, blobs rapidly became continuous in nature as evidenced from test pit observations.

Defining the shape and size of blobs presents another problem. Observations from this project indicated that the majority of the randomly viewed subsamples contained blobs that appear to be oblong in dimension. In samples containing more than 4% Bunker C oil the length to width ratio seemed much greater than in soils containing less oil. Discernable blobs in diesel affected samples containing less than 1000 mg/kg TPH appeared more cylindrical than oblong. These observations are for samples with similar grain size distributions. The specific reasons for these variations were not evaluated.

The complexity of mass transfer and flow of petroleum are substantial. It is evident that the literature and practice alike present conflicting results and theories. Nevertheless, now that mathematical expressions for slow dissolution have been identified, an attempt can now be made to estimate that amount of hydrocarbon amenable to hydraulic sweeping.

HEAVY OIL RECOVERY

At the Dunsmuir Yard, the dominant hydrocarbon fraction exceeds C_{20}. Three major driving forces influence the amount of oil that can be recovered without additives: hydraulic forces, capillary forces, and chemical (mass) forces. Chemical forces may be described in terms of mass concentration gradients limiting the flux rate from a given control volume (Equation 2). Pumping wells or collection trenches installed downgradient from the heavy oil source areas induce an artificial hydraulic gradient in the aquifer. Oil in both pure and dissolved phases are then drawn toward the hydraulic sink and recovered.

Pressure Gradients

Capillary force is an important interaction mechanism which is measured by the capillary pressure. Mathematically, it is the pressure difference between the NAPL and water phases, and is described using the Young and Laplace relation. The simplest geometry of a NAPL blob is hemispherical. Capillary pressure using hemispherical shapes is given by:

$$p^{cap} = p^{NAPL} - p^{aq} = \frac{2Y_{ow}}{r_{1,2}}$$

Here, p^{cap} is the capillary pressure, p^{NAPL} is the pressure in the NAPL , p^{aq} is the pressure in the aqueous phase, Y_{ow} is the NAPL/water interfacial tension, and $r_{1,2}$ are the radii of curvature of the hemispherical interface (blob).

The Young and Laplace equation can be used to estimate the capillary forces that hold NAPL drops trapped in the sediment matrix, and the hydraulic gradient required to force the NAPL through particle constrictions. If $P_1{}^{aq}$ and $P_2{}^{aq}$ are the aqueous phase pressures on the up and downgradient sides of the blob, and r_1 and r_2 are the radii of curvature for the hemispherical blob, the hydraulic gradient necessary to push the droplet through the construction, /P is then given by:

$$\Delta P = P_1{}^{aq} - P_z{}^{aq} = P_2{}^{cap} - P_1{}^{cap} = 2Y_{ow} \left\{ \frac{1}{r_2} - \frac{1}{r_1} \right\}$$

Table 3-2 presents properties of n-alkanes which are available in the literature. Using pentadecane to represent the heavy hydrocarbon mixture,(it was the largest hydrocarbon for which sufficient data was available), the applicable interfacial tension is 58 dyne/cm. Next we apply the work of Berg (1975) who provides a relationship for estimating pore throat diameters based on rhombohedral packing of spheres which is helpful in selecting r_1 and r_2 values. Pore radii values less than 5 μm correspond to silts, 5 to 30 μm correspond to fine sand, and 30 to 200um correspond to sands and gravel.

Using these values, the pressure gradient needed to push a NAPL blob through the unsaturated soils can be estimated. Figure 3-3 shows the results as a function of hydraulic gradient and droplet size for a wide range of gradients and pore radii. The pressure drop needed to move the NAPL is many orders-of-magnitude larger than the typical natural gradient of 0.010 to 0.25 kPa/m that is observed across the site during any season. Even when r_2 and r_1 are minor as they are when considering the trapped diesel fraction, the required pressure drop is still very large compared to the typical natural gradients in site aquifer. Thus, these calculations suggest that natural gradients in even this porous rivertine environment are insufficient to mobilize NAPL that is trapped in discrete units. Even continuous NAPL under low head conditions would be subject to similar physical constraints.

Table 3-2 **Properties of n-alkanes Applied to Transport Analyses**

Compound	Solubility[1] (mg/l)	Interfacial Tension (dyne/cm)	Density (g/cm^3)	Kinematic Viscosity (cP)	Diffusion Coefficient (cm^2/s)	log K_{ow}	F_{oc} (%)	Kd^2
n-octane	0.431	50.8	0.702	0.412	2×10^{-8}	5.18	1.2	1.1×10^3
n-nonane	0.122	51.0	0.718	0.474	1×10^{-8}	5.67	1.2	3.5×10^3
n-decane	0.022	53.2	0.730	0.556	6×10^{-9}	6.69	1.2	3.6×10^4
n-pentadecane	0.001	58.0	0.7685	1.116	1×10^{-9}	8.54	1.2	4.3×10^4
n-eicosane	<0.0005	>60.0	0.7887	1.917	3×10^{-10}	>10.0	1.2	4.9×10^4

Compound	K_H (atm-m^3/mol)	MW	Vapor Pressure (mmHg)	log K_{oc}
n-octane	3.22	114.23	11	4.97
n-nonane	5.95	128.26	3.22	5.46
n-decane	1.87×10^{-1}	142.28	1.35	6.48
n-pentadecane	38.2	223.8	0.12	6.55
n-eicosane	—	285.6	<0.01	6.69

[1] Values at 25°C.

[2] Actual value may vary by over one order of magnitude.

Sources: API, 1982
  API, 1953
  Hoffman, 1969

It should be noted that large variations of more than four orders-of-magnitude in grain size are evident in the Dunsmuir rivertine sediments. As such, grain size distribution plays a significant role in actual transportability of the heavy oil. Although simple in its structure, this analysis brings into focus the magnitude of capillary forces acting in the subsurface.

The foregoing discussion emphasizes two unique fate dynamics in effect at the Dunsmuir Yard: hydraulic control and diffusion. Recovery of crude oil resident above residual saturation, considered here to be one percent (10,000 mg/kg), will be a dominant function of hydraulics as expressed in terms of capillary pressure. The surface area encompassed by the Bunker C oil is approximately 8,600 yd^2, or about 26% of the total surface area containing detectable TPH in sediments. The balance, approximately 24,500 yd^2, largely is well below a concentration considered to be mobile or a separate phase. This fraction will contribute migrating hydrocarbons through fickian diffusion into mobile (bulk fluid) water at trace concentration over an extended period of time. Assessment of mass removal must be viewed in the context of reducing the most concentrated areas. Trace TPH migration from other areas can be evaluated only generally via diffusion processes which are, in turn, limited by a compound's solubility in water.

Hydraulic Removal

The capillary forces (pressures) which resist mobilization can be overcome by viscous forces associated with hydraulic gradient or by buoyancy forces. Residual Bunker C oil at the Dunsmuir Yard is subjected to relatively weak buoyancy forces due to density differences. Capillary and viscous forces are more significant than buoyancy forces with regard to mobilization of residual oil in pore spaces in the saturated zone.

The dimensionless ratio of capillary forces to viscous forces is known as the capillary number (Nc):

$$N_c = \frac{k \cdot \rho_w \cdot g \cdot J}{\sigma}$$

where

k	= intrinsic permeability of porous medium (cm^2)
ρ_w	= the density of water (g/cm^3)
σ	= interfacial tension (dyne/cm)
g	= gravitational acceleration (cm/sec^2)
J	= hydraulic gradient (cm/cm)

Experimental studies by Wilson and Conrad (1984) established a strong correlation between displacement of residual liquids and capillary number, when the hydraulic gradient was greater than the critical value needed to initiate motion of some of the residual liquid blobs or droplets. This critical value of the capillary number depends on residual saturation.

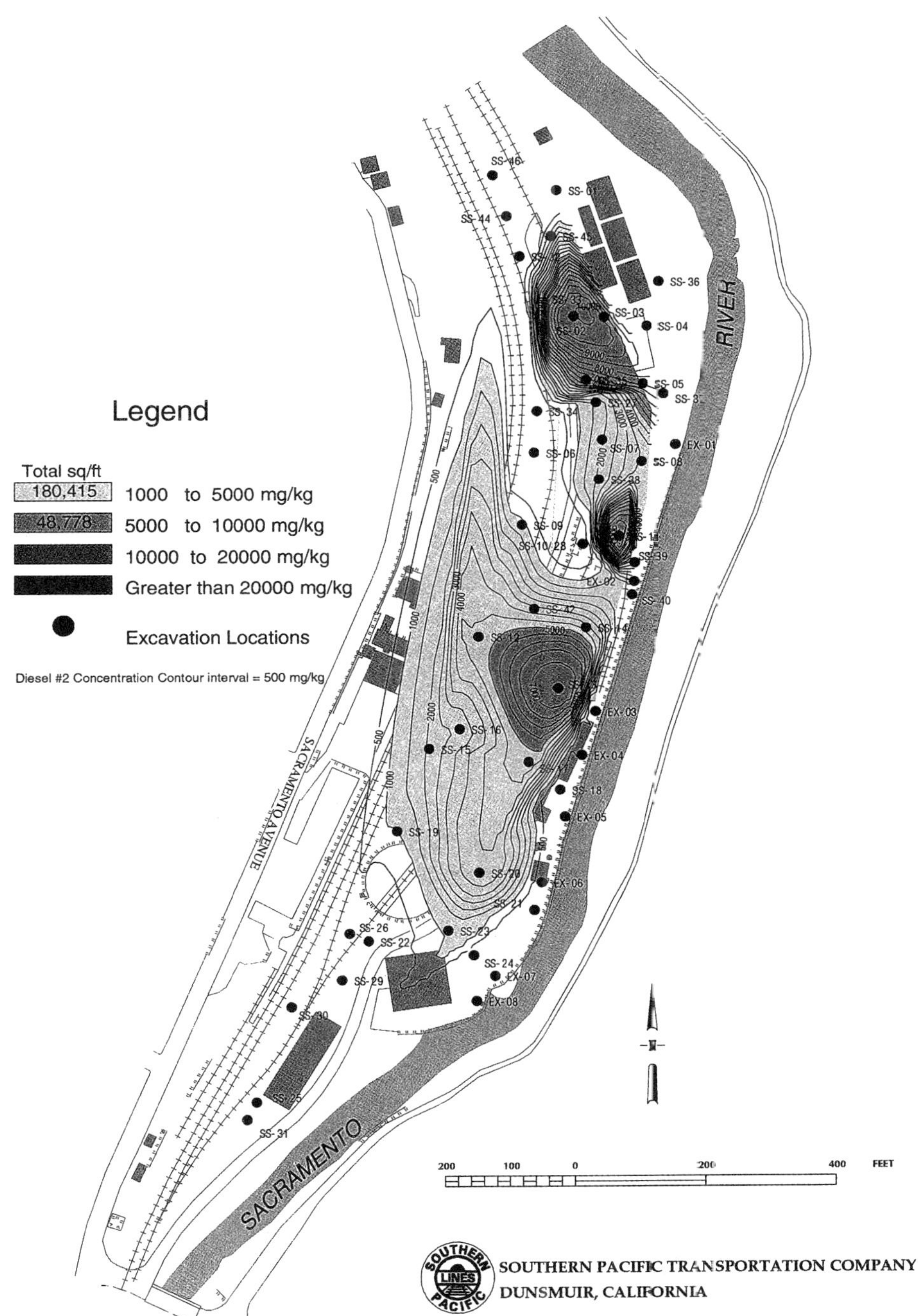

Figure 3-3   Pressure Gradient Required to Overcome Capillary Forces

Liquid solute saturation is the volume of liquid contaminant per unit void volume:

$$S_o = V_o / V_v$$

where

S_o = organic or liquid contaminant saturation

V_o = volume of hydrocarbons

V_v = volume of voids

As the liquid contaminant is displaced, its saturation is reduced to a residual:

$$S_v = V_{do} / V_v$$

where

S_{or} = residual liquid contaminant saturation

V_{do} = volume of discontinuous liquid contaminant

V_v = volume of voids

The relationship between the capillary number and normalized residual saturation for site conditions and a reference bead pack case is shown in Figure 3-4; (S_{or}final residual saturation, S_{or}^*,initial saturation). Nc^* is the critical capillary number at which motion starts. Nc^{**} denotes the capillary number necessary to displace all of the residual hydrocarbon, in theory.

Some of the liquid hydrocarbon will be removed when $Nc > Nc^*$ and a large percentage can be removed when $Nc > Nc^{**}$. Figure 3-5 is a plot of hydraulic gradient of the water gradient (J) versus intrinsic permeability (k) taken from Wilson and Conrad (1984) and modified to express Dunsmuir site conditions. The capillary number of Equation 6 for the critical value Nc^* and for various hydrocarbon water interfacial tensions. This plot describes hydraulic gradients necessary to initiate blob mobilization and are similar in form to the pressure-related analysis provided in Figure 3-4.

Upon review of Figure 3-5, given an average permeability of 2.5×10^{-6} cm^2 in site sediments, a hydraulic gradient of 0.85 would be required to mobilize mid-range n-alkanes (C_8-C_{19}) towards a collection system. To accomplish this, a collection system would be needed that could increase the natural gradient by a factor of 42, from 0.02 to 0.85. A gradient of this magnitude is marginally achievable in typical alluvial materials; however, at the site such a gradient is only possible within 10 feet of a fully depressed pumping well. It can also be seen that to extract most of the mid-range alkanes, a hydraulic gradient above 100 may be necessary, which is two orders-of-magnitude above a practical operating range. Mobilization of even longer chain alkanes (C_{20}-C_{35}) which dominate the three concentrated zones does not appear to be achievable by simple hydraulics alone.

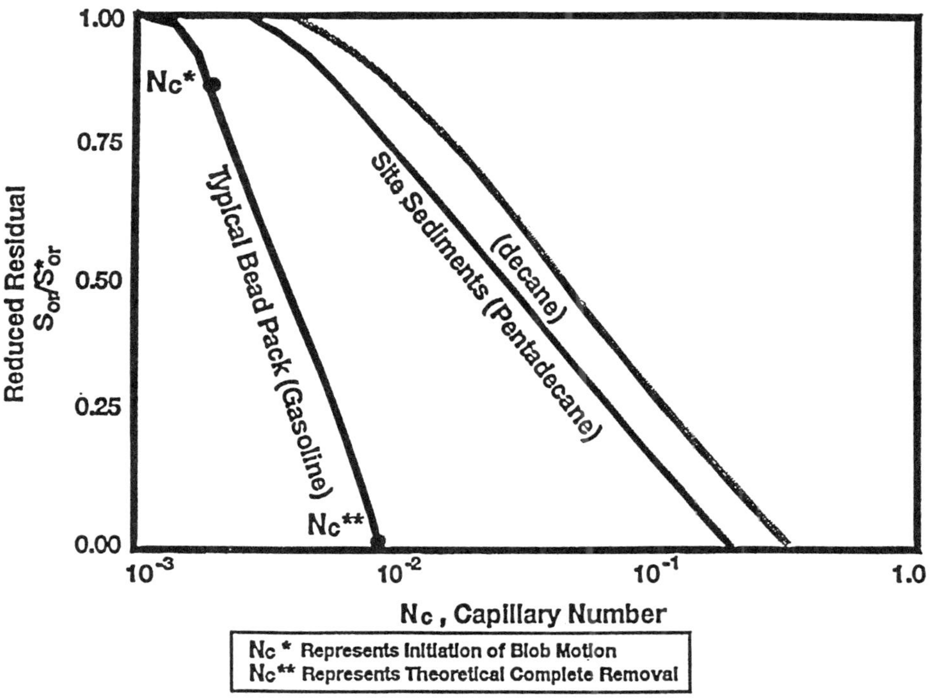

Figure 3-4   Oil Removal as a Function of Saturation and Capillary Pressure

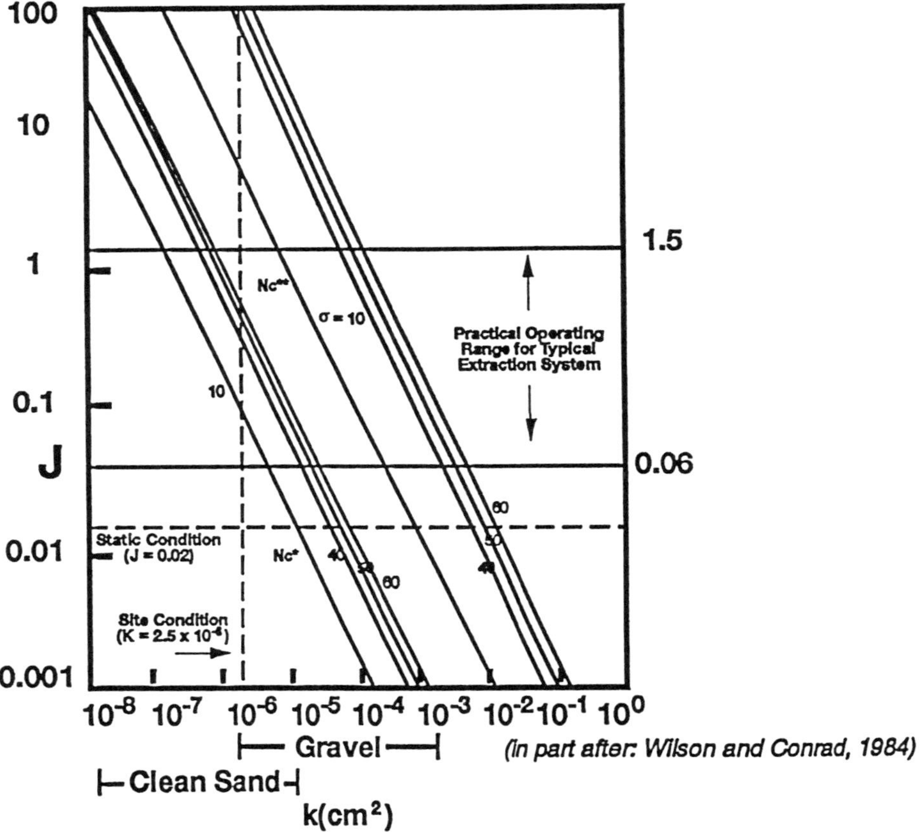

Figure 3-5   Hydraulic Gradient Required for Weathered Oil Removal

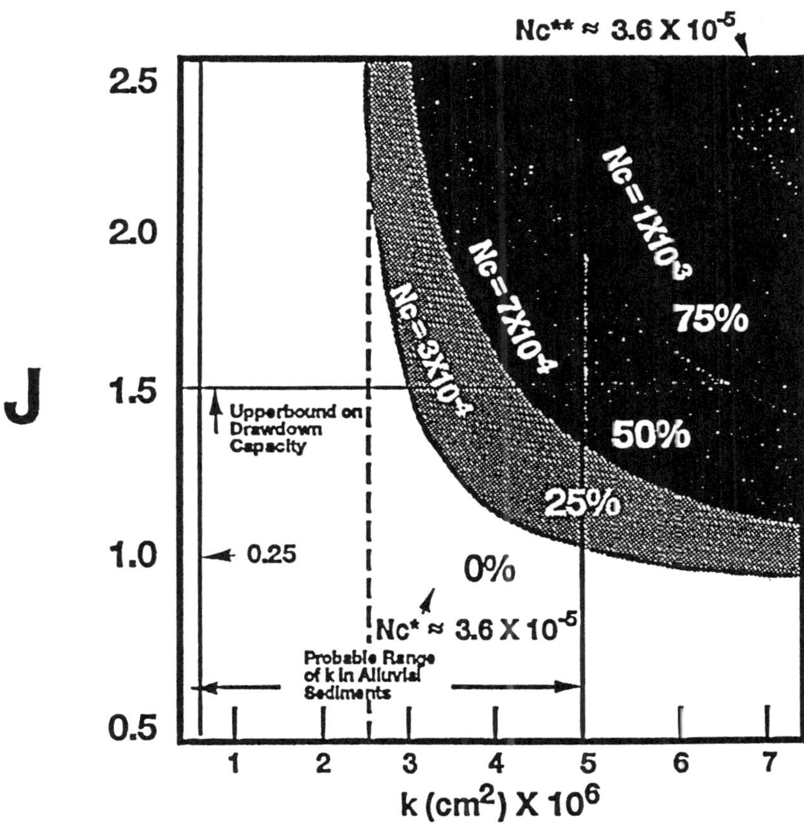

Figure 3-6   Estimating Pentadecane Removal Expressed as a Function of Varying Capillary Pressures in the Subsurface

Figure 3-6 illustrates residual pentadecane (σ = 58 dyne/cm) recovered at various capillary pressures and superimposes Dunsmuir site conditions as a function of gradient and permeability. A range of intrinsic permeabilities across the Site were applied to Equation 6 and compared to pentadecane mobilization at varying hydraulic gradients. For the case of pentadecane, one could expect to remove, on average, less than 25% of the mass by hydraulic gradient inducement alone under the maximum permeability and gradient. The plausible range for pentadecane mobilization lies between 0 and 25%. Thus hydraulic sweeping even longer chain n-alkanes would be totally ineffective.

CONCLUSION

In summary, residual Bunker C oil probably cannot be removed from the subsurface at the Dunsmuir Yard by any type of hydraulic extraction system alone. It is easier to mobilize hydrocarbons with lower interfacial tensions in water. Induced hydraulic gradients must be combined with methods to reduce interfacial tensions that will allow for hydrocarbon mobilization.

Several other salient points can be gleaned from the data collected at the Dunsmuir Yard. First, equilibrium partitioning greatly overestimates ground water concentrations at the site. Second, toxic components of petroleum, BTEX and short-chain aromatics resulting from historical spills, have advected out of the site subsurface, leaving the more viscous and less toxic fraction trapped in interstitial pore spaces. Finally, trace amounts of long-chain n-alkanes and aromatics will continue to dissolve from the residual mass for several decades without appreciably influencing the amount of hydrocarbon mass resident in site sediments.

REFERENCES

1.   Abriola, L.M. and G.F. Pinder, 1985. A Multiphase Approach to the Modeling of Porous Media Contamination by Organic Compounds I. Equation Development, *Water Resources Research*, V. 21, #1, pp. 11-18.

2.   American Petroleum Institute (API), 1953.   Selected Values of Physical and Thermodynamic Properties of Hydrocarbons and Related Compounds, Carnegie Press, Pittsburgh, PA. pp. 268-336.

3.   American Petroleum Institute (API), 1972. The Migration of Petroleum Products in Soil and Ground Water, Principles and countermeasures, technical report ,35pp., Am. Petrol. Inst., Washington, DC.

4.   American Petroleum Institute (API), 1982. API Technical Data Book, Chapter 10, Interfacial Tension, Chapter 13, Viscosity. 872 pp.

5.   Anderson, D., K.W. Brown, and Jan Green, 1982. Effect of Organic Fluids on the Permeability of Clay Soil Liners, Proceedings of the Eighth Annual Research Symposium on Land Disposal of Hazardous Waste, in EPA 600/9-82-002 ,pp.179- 190.

6.   Anderson, M.R., R.L. Johnson, and J.F. Pankow, 1987. The Dissolution of Residual Dense Nonaqueous Phase Liquid (DNAPL) from a Saturated Porous Medium,Proc. NWWA Conf. on Petroleum Hydrocarbons and Organic Chemicals in Groundwater, pp. 409-428.

7.   Atwater, J.W., 1984. A Case Study of a Chemical Spill: Polychlorinated Biphenyls Revisited, *Water Resour. Res.*, 20(2), pp. 317-319.

8.   Baehr, A., 1987. Selective Transport of Hydrocarbons in the Unsaturated Zone Due to Aqueous and Vapor Phase Partitioning, *Water Resour. Res.*, 23(10), pp.1926-1938.

9.   Bear, J., 1979. Hydraulics of Groundwater, McGraw-Hill, New York, New York.

10.   Berg, R., 1975. Capillary Pressures in Stratigraphic Traps, *Am. Assoc . Pet. Geol. Bull.*, 59(6), pp. 939-956.

11.   Bowman, C.W., D.M. Ward, A.I. Johnson, and O. Trass, 1961. Mass Transfer from Fluid and Solid Spheres at Low Reynolds Numbers, *Can. J. Chem. Eng.*, 39(1),pp. 9-13.

12.   Brusseau, M.L., and P.S.C. Rao, 1989. Sorption Nonideality During Organic Contaminant Transport in Porous Media, *CRC Crit. Rev. Environ. Control*, 19(1), pp.33-99.

13.   Cohen, R.M., R.R. Rabold, C.R. Faust, S.O. Rumbaugh, III, and J.R. Bridge, 1987. Investigation and Hydraulic Containment of Chemical Contamination: FourLandfills in Niagara Falls, Civil Engineering Practice, *J. Boston Soc. Civ. Eng.*2(1), pp. 33- 58.

14.   Conrad, S.H., J.L. Wilson, W. Mason, and W. Peplinski, 1989. Observing the Transport and Fate of Petroleum Hydrocarbons in Soils and in Groundwater Using Flow Visualization Techniques, in Environmental Concerns in the Petroleum Industry, S.M. Testa ed., pp. 1-13.

15.   Corapcioglu, M.Y. and A.L. Baehr, 1987. A Compositional Multiphase Model for Groundwater Contamination by Petroleum Products I. Theoretical Considerations, *Water Resources Research*, Vol. 23, No. 1, pp. 191-200.

16. deZabala, E.F., and C.J. Radke, 1986. A Non-Equilibrium Description of Alkaline Water Flooding, *Soc. Petrol. Eng. J.*, 26(1), pp. 29-43.

17. Dietz, D.N., 1980. The Intrusion of Polluted Water into a Groundwater Body and the Biodegradation of a Pollutant, Proceedings of the National Conference on Control of Hazardous Material Spills, U.S. EPA.

18. Dorgarten, H.W. and C.F. Tsang, 1990. Three Phase Simulation of Organic Contaminants in Aquifer Systems, Conference on Subsurface Contamination by Immiscible Fluids, Internation. Assoc. of Hydrogeologists, Calgary, Alberta.

19. Feenstra, S. and J. Coburn, 1986. Subsurface Contamination from Spills of Denser Than Water Chlorinated Solvents, Calif. WPCA Bull., 23(4), pp. 26-34.

20. Fried, J.J., P. Muntzer, and L.Zilliox, 1979. Groundwater Pollution by Transfer of Oil Hydrocarbons, *Groundwater*, 17(6), pp. 586-594.

21. Friedlander, S.K., 1957. Mass and Heat Transfer to Single Spheres and Cylinders and Low Reynolds Numbers, *Aiche J.*, 3(1), pp. 43-48.

22. Geller, J.T. and J.R. Hunt, 1989. Non-aqueous Phase Organic Liquids in the Subsurface: Dissolution Kinetics in the Saturated Zone, paper presented at the International Symposium on Processes Governing the Fate of Contaminants in the Subsurface Environment, Int. Assoc. of Water Pollut. Res. and Control, Stanford, Calif.,July 23-26.

23. Goltz, M.N., and P.V. Roberts, 1987. Using the Method of Moments to Analyze Three-Dimensional Diffusion-Limited Solute Transport from Temporal and Spatial Perspectives, *Water Resour. Res.*, 23(8), pp. 1575-1585.

24. Hoag, G. and M. Marley, 1986. Gasoline Residual Saturation in Unsaturated Uniform Aquifer Materials, *J. Env. Eng.*, 112(3), pp. 586-604.

25. Hoffman, B., 1970. Dispersion of Soluble Hydrocarbons in Groundwater Streams, in Advances in Water Pollution Research, Vol. 2, HA1-7, International Association on Water Pollution Research, Pergamon, New York.

26. Hunt, J.R., N. Sittar, and K.S. Udell, 1988. Non-Aqueous Phase Liquid Transport and Cleanup, 1, Analysis of mechanisms, *Water Resour. Res.*, 24(8), pp. 1247-1258.

27. McKee, J.E., F.B. Laverty, and R.M. Hertel, 1972. Gasoline in Groundwater, *J. Water Pollut. Control Fed.*, 44(2), pp. 293-302.

28. Melrose, J.C. and C.F. Brander, 1974. Role of Capillary Forces in Determining Displacement Efficiency for Oil Recovery by Water Flooding, *J. Can .Pet. Technol.*, Vol. 13.

29. Moench, A.F. and A. Ogata, 1981. A Numerical Inversion of the LaPlace Transform Solution to Radial Dispersion in the Porous Medium, *Water Resources Research, V. 17, #1.*

30. Ng, K.M., H.T. Davis, and L.E. Scriven, 1978. Visualization of Blob Mechanism in Flow Through Porous Media, *Chem. Eng. Sci. 33*, pp. 1009-1017.

31. Ogata, A., 1958. Dispersion in Porous Media, Ph.D. dissertation, Northwestern Univ., Evanston, Ill.

32.   Pfannkuch, H.O., 1984. Determination of the Contaminant Source Strength from Mass Exchange Processes at the Petroleum-Groundwater Interface in Shallow Aquifer Systems, in Proceedings of the NWWA Conference on Petroleum Hydrocarbons and Organic Chemicals in Groundwater, pp. 111-129, National Well Water Association, Dublin, Ohio.

33.   Schwille, F., 1984. Migration of Organic Fluids Immiscible with Water in the Unsaturated Zone, in Pollutants in Porous Media, Ed. B. Yaran, D. Dugan, and J. Goldschmid, Springer-Verlag, New York.

34.   Schwille, F., 1988. Dense Chlorinated Solvents in Porous and Fractured Media, translated by J.F. Pankow, Lewis Publishers, Chelsea, Michigan, p. 146.

35.   Valocchi, A.J., 1986. Effect of Radial Flow on Deviations from Local Equilibrium During Sorbing Solute Transport Through Homogeneous Soils, *Water Resour. Res.*, 22(12), pp. 1693-1701.

36.   van Genuchten, M.Th., and P.J. Wierenga, 1976. Mass Transfer Studies in Sorbing Porous Media, I, Analytical Solutions, *Soil Sci. Soc. Am. J.*, 40(4), pp. 475-480.

37.   Welty, J.R., C.E. Wicks, and R.E. Wilson, 1969. Fundamentals of Momentum, Heat and Mass Transfer, John Wiley, New York.

38.   Wilson, J.L., and S.H. Conrad, 1984. Is Physical Displacement of Residual Hydrocarbons a Realistic Possibility in Aquifer Restoration?, in Proceedings of the NWWA Conference on Petroleum Hydrocarbons and Organic Chemicals in the Subsurface, pp. 274-298, National Well Water Association, Dublin, Ohio.

39.   Zilliox, L., P. Montzer, and J.J. Menanteau, 1973. Probléme de l'echange entre un produit pétrolier immobile et l'eau en mouvement dans un milieu po reax.*Revue IFP.* 28(2), pp. 185-200.

CHAPTER 4

Soil Remediation in Railroad Yards — How Clean is Clean?

Marleen A. Troy, Ph.D. and James L. Brown, Ph.D., CPSS
OHM Remediation Services Corp — Trenton, New Jersey

At most railroad facilities there are many challenges to the remediation of contaminated soils. The most obvious challenge is delineating what areas are truly contaminated, and what levels are considered 'background'. Most railroad yards are located in heavy industrial areas, and have been in use for decades. The manner in which the railroads were constructed, as well as how they were operated needs to be understood when evaluating remediation cleanup criteria. Soil remediation at railroad facilities lends credence to the premise that generic cleanup criteria are not appropriate for all sites. Current and future land use needs to be considered when establishing soil cleanup guidelines.

The importance of establishing background soil conditions prior to negotiating soil cleanup criteria with state regulators cannot be overestimated. Unfortunately site cleanups often proceed without either a clear statement of cleanup objectives or recognition of background levels of total petroleum hydrocarbons (TPH) and polycyclic aromatic hydrocarbons (PAHs). It is quite possible that background contaminant levels may by themselves exceed soil cleanup criteria. Railroad yards are usually located in areas where fossil fuels have been burned for decades. Direct deposition from the atmosphere has caused surface soils in these areas to contain high levels of several PAHs. Some of these compounds such as benzo-a-pyrene are considered carcinogenic, and surface soil cleanup standards can be quite low when these are present.

Selection of mutually acceptable background soil sampling locations between a railroad company and state regulatory agency can be very difficult. Areas within railroad yards not impacted by even minor petroleum spills or deposition from adjacent industrial activities are quite rare, or very difficult to identify with historical aerial photography. The typical railroad yard is also a heterogeneous mixture of fill materials, many of unknown origin. Figure 4-1 is an illustration of a typical railroad bed cross section. The railroad roadbed is the regular prepared subgrade on which the ballast section, ties, and rails are placed.[1] The roadbed shoulders typically extend at a minimum 18 inches beyond the toe of the ballast slope to provide adequate support to the ballast section. Ideally, sandy or gravelly material was placed on the upper portions of the fill material used to comprise the subgrade. A clay of low plasticity was rolled to maximum field density in order to act as an impervious waterproofing layer. It was

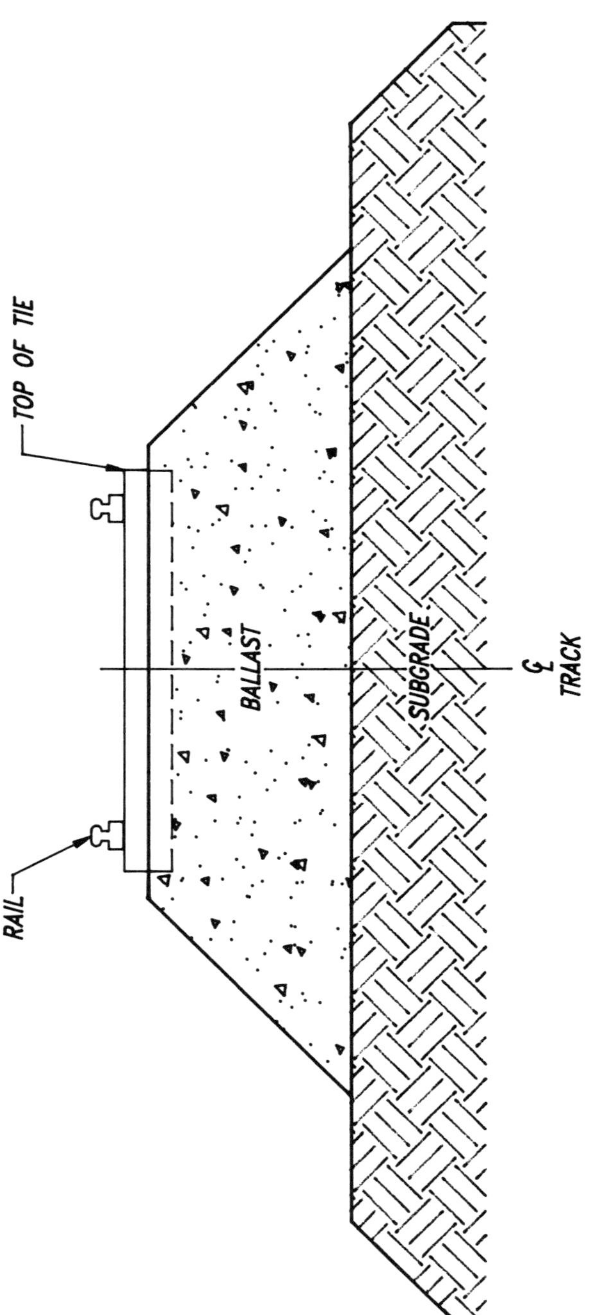

Figure 4-1 Typical railroad roadbed cross section (After Hay, 1953).

desirable to use better soil materials as the base and slope coverings, with clays and silts being used for the interior of the fill. Ballast is gravel or broken stone laid on a railroad bed. The kinds of ballast that may have been or are currently in use at railroad yards are crushed stone, crushed slag, prepared gravel, pit run gravel, chat (the tailings or refuse from zinc, lead and silver mining), cinders, chert (compact, flint-like siliceous rock), sand, burned coal mine tailings (caliche), and sea shells.[1] Homogeneous soil conditions at railroad yards are therefore the exception, not the rule. Fill materials that contain coal or shale residues are known to cause positive matrix interferences for both TPH and certain PAH compounds. The problem of matrix interferences and their impact in achieving a successful cleanup is presented later in this chapter.

If after a thorough historical record search and site inspection, acceptable background soil sampling locations cannot be found onsite, then an offsite location should be considered. More than one (1) background sampling location is advisable. When an acceptable background sampling area has been selected, a large enough number of soil samples should be collected to establish a statistical basis for comparison with petroleum impacted soils. At least five (5) or six (6) sets of separate surface and subsurface soil samples should be collected. Typically, background data are averaged, and an upper confidence limit is established for the estimated true mean value of the sample population. This upper confidence level value can then be compared to individual data to determine whether the levels of individual contaminants exceed the background level. Wherever background contaminant levels exceed cleanup standards, site cleanup to background conditions will usually be required.

The most commonly spilled hazardous material in railroad yards is diesel fuel. State cleanup guidelines or regulatory standards for total petroleum hydrocarbons (TPH) for diesel fuel spills in soil vary widely among northeastern states where most railroad yards are located. Guidelines expressed as TPH range from 10 to 10,000 parts per million (ppm). Most state guidelines for TPH are below 100 ppm. In general, these numerical guidelines are not risk-based, and are very conservative. A recent comprehensive review of human health risks from diesel fuel-impacted soil concluded that soil cleanup levels ranging from approximately 1,100 to 11,000 ppm TPH would be protective of human health.[2] These levels represent an increased cancer risk of 1 x 10^{-5} for a residential land use scenario. This risk level is in the middle range of risks deemed acceptable by the United States Environmental Protection Agency (U.S. EPA). It seems clear soil cleanup levels of 100 ppm TPH or lower represent risk levels that can be orders of magnitude more conservative than U.S. EPA guidelines for both CERCLA and RCRA sites.

The variability of soil cleanup guidelines for diesel-fuel impacted soils in states throughout the northeast is typical of that found nationally. In Massachusetts, cleanup standards have not been promulgated. All decisions on 'How Clean is Clean?' are currently determined via a site-specific risk evaluation. Some states have developed cleanup criteria based on specific target compounds, rather than indicators of total organic contaminants like TPH. In Pennsylvania, for example, soil cleanup criteria for virgin fuel spills specify compliance levels for benzene, toluene, ethylbenzene, and total xylenes (BTEX) in addition to TPH. Soil cleanup levels in Pennsylvania are based primarily on protection of groundwater quality. Delaware has developed a similar system of soil cleanup criteria based primarily on groundwater protection scenarios. In New York, the Toxicity Characteristic Leaching Procedure (TCLP) has been accepted as a method of determining leachability of target compounds in petroleum contaminated soil. Soil cleanup guidance values in New York are also based primarily on protection of groundwater quality. Compliance requires that soil TCLP extracts be less than or equal to maximum contaminant levels (MCLs) for drinking water. An alternative guidance value can be selected for the total contaminant level present, which cannot exceed 20 times the MCL. This alternative is based on the assumption that 100 percent of the contaminant is leachable. New Hampshire has developed soil cleanup criteria that consider total BTEX levels, but do not differentiate between these individual volatile compounds, even though federal drinking water MCLs vary considerably for these compounds. In Vermont, soil action levels are generally 20 times the state drinking water standards for specific contaminants.

New Jersey has yet to promulgate regulatory soil cleanup standards. In the interim, soil cleanup guidelines are being used for most cleanups of diesel fuel impacted soil. New Jersey has proposed establishment of both surface (0 to 2 feet) and subsurface soil cleanup guidelines, based primarily on direct exposure and potential for groundwater contamination, respectively. New Jersey has also proposed guidelines based on land use by differentiating between residential and industrial sites. For example, a level of total organic contamination at a residential site cannot exceed 1,000 ppm. At industrial sites, the comparable level is 10,000 ppm. In general, if TPH levels are less than 1,000 ppm, no cleanup is required. When TPH levels exceed 10,000 ppm, cleanup action is mandatory. Selection of soil cleanup levels based on an industrial land use classification requires adoption of institutional control such as permanent deed restriction. Recognition of both current and future land use in setting soil cleanup standards has a definite advantage when considering the cost of cleanup. If the basis for site cleanups is arbitrarily derived, very conservative cleanup standards, the cost of cleanup can be prohibitive. Other land use scenarios should also be considered when determining cleanup standards. If commercial development is proposed following site cleanup, the type of development should definitely be considered when selecting soil cleanup goals. For example, the potential for human exposure by direct contact with diesel fuel-impacted soil can be greatly reduced if the site is used for building construction or is paved.

Another issue related to 'How Clean is Clean?' is to determine what constitutes an acceptable cleanup. This is usually determined by collection and analysis of confirmatory soil samples, followed by comparison with a cleanup criterion. Under ideal conditions, a sufficient number of soil samples is collected to establish a statistical confidence level associated with a clean soil boundary. This is seldom done in practice, because it requires the collection of a large number of soil samples, which is usually cost prohibitive. Given the existence of cleanup guidelines, regulatory agencies have an obligation to develop some process of determining compliance with cleanup criteria. Without such a process, if only one of several confirmatory soil samples exceeds a cleanup criterion, the entire cleanup could be judged unacceptable. A current example from a northeastern state illustrates this problem. A total of six (6) confirmatory soils samples were collected to document the quality of a bioremediation landfarm cleanup of TPH contaminated soil. TCLP extracts of soil samples were analyzed for volatile organic compounds (VOCs). Soil cleanup guidelines required compliance of TCLP soil extracts with state drinking water standards. The applicable state drinking water standard for benzene, the contaminant of greatest concern, was 0.0007 ppm. Five (5) of the six (6) TCLP extracts contained less than 0.0007 ppm benzene; a sixth sample contained just 0.004 ppm. Because only one (1) of six (6) samples exceeded cleanup guidelines, the cleanup was considered incomplete.

One example of a process to determine compliance with cleanup standards is from New Jersey, which has proposed soil cleanup standards for contaminated sites. In the proposed regulations, compliance with a soil cleanup standard is achieved when:

- The arithmetic mean of the contaminant concentration in all soil samples in an area of concern is less than or equal to the applicable soil cleanup standard for that contaminant, and;

- No single soil sample exceeds the applicable soil standard by a factor of more than:

  - ten for a standard less than or equal to 10 ppm;

  - five for a standard greater than 10 ppm, but less than or equal to 100 ppm;

  - two for a soil standard greater than 100 ppm; and

- No more than 10 percent of soil samples exceed the applicable cleanup standard (or one sample if 2 to 10 samples are collected).

Regardless of the details of what constitutes an acceptable cleanup, it is important that some process exists to determine compliance with regulatory cleanup standards.

Because soil-borne contaminants frequently impact groundwater, the approach to groundwater remediation is also linked to remediation of soil. As cited above, soil cleanup guidelines are frequently established for the protection of groundwater quality. Because most railroad yards are located in highly developed, industrial areas, groundwater is seldom used for potable supplies. In spite of this, groundwater cleanup goals often require attainment of drinking water standards. This may be technically unfeasible as well as cost prohibitive. In New Jersey, an alternative approach to groundwater remediation has been proposed that combines active remediation with natural remediation processes to maximize effective use of finite resources for site cleanup. This approach allows natural remediation of contaminant plumes when no-long term unacceptable risk to human health or the environment is expected. The approach is based on the premise that contaminants in the groundwater will chemically or biochemically degrade, and are subject to attenuation by sorption. As a result, contaminant concentrations reaching a receptor can be reduced by natural processes to levels below applicable cleanup standards.

The primary criterion for proposing a natural remediation compliance alternative is evidence that contaminants on site will degrade or naturally attenuate. There is sufficient evidence in the literature to indicate that most contaminants, including chlorinated solvents are either subject to degradation or sorption or both. The contaminants of greatest concern in diesel fuel for their potential impact to groundwater are benzene, toluene ethylbenzene and the xylenes (BTEX). Of these, benzene is the most critical. It is the only BTEX compound either known or suspected of being a carcinogen. As a result, groundwater quality standards for benzene are the most stringent, usually one (1) part per billion or less. Unfortunately, benzene is usually found in groundwater following most diesel fuel spills in railroad yards.

There are several requirements for acceptance of a natural remediation alternative. Although the following requirements are specific to proposed regulations in New Jersey, similar requirements can be expected whenever this process is considered as an alternative to active remediation. The proposed requirements for New Jersey are as follows:

- Supply evidence that all sources of contamination and free product have been controlled, and that contaminants present are not expected to migrate to a human or ecological receptor above applicable standards;

- Provide documentation regarding the contaminant's degradability or potential for attenuation;

- Identify and discuss site-specific characteristics that indicate conditions are favorable for degradation and/or attenuation;

- Establish a groundwater monitoring program to evaluate to adequacy of source control, to track contaminant concentration over time, and to monitor contaminant degradation and attenuation within and downgradient of the plume;

- Establish a sentinel well system designed to detect contamination in groundwater prior to reaching any potential receptor. The well must be located at least one years's travel time upgradient of the receptor;

- Provide written documentation of current and potential groundwater uses based on a 25-year planning horizon;

- Provide written notices to all property owners and occupants of property under which contaminated groundwater is expected to migrate;

- Provide evidence that all access agreements needed to monitor groundwater quality have been or can be obtained.

It is obvious that these proposed requirements for compliance are extensive. But given these requirements, a cleanup plan which relies on natural remediation may still be preferable to active remediation of groundwater. Active processes such as pump-and-treat scenarios are very costly, and often do not provide assurance that contaminants can be reduced to acceptable levels within a reasonable time frame. We believe a natural remediation alternative is well suited to most railroad sites because they are usually located in industrial areas where groundwater is not typically used for potable water supplies.

A successful cleanup of diesel fuel-impacted soil and groundwater may also require compliance with groundwater quality standards. If groundwater quality standards are developed for protection of aquifers for potable water supplies, then federal and state drinking water standards will usually apply. Such standards are by definition very conservative. Most railroad yards are located in industrial areas where use of groundwater for potable supplies is very limited or nonexistent, and where there has been a long history of industrial pollutant discharge. In these areas, groundwater remediation goals should reflect both background water quality and actual or projected groundwater use.

New Jersey recently adopted groundwater quality standards that recognize the existence of heavily polluted groundwaters. In regulations adopted this year, groundwater is divided into three general classes established for the protection of designated uses of groundwater:

- Class I is groundwater of special ecological significance;

- Class IIA is groundwater suitable for use for potable water supply at current levels of water quality with conventional treatment;

- Class IIB is groundwater suitable for potable supply only if existing poor quality groundwater which has been damaged by pollution can be enhanced or restored for potable supply use with conventional treatment;

- Class III is groundwater with uses other than potable water supply.

The concept of designating Class IIB groundwater was well received by industry. It recognizes the existence of widespread areas in which Class IIA criteria are exceeded, and in which there is little or no current or potential groundwater use. It is also consistent with protection of groundwater supplies, since even relatively limited use areas for water supply retain the IIA classification. Groundwater cleanup remedies for Class IIB groundwaters would be required to achieve source control, free product removal and protection of downgradient receptors. However, the existence of a Class IIB groundwater would not be a license for use as dumping sites for hazardous substances. Use of a natural remediation alternative as described above would be very appropriate for Class IIB groundwaters.

In practice, the designation of Class IIB groundwaters has not yet facilitated groundwater cleanup in railroad yards and industrial areas where this classification is most appropriate. This is because at present there are no delineated Class IIB areas. The New Jersey Department of Environmental Protection (NJDEP), because of limited resources, has yet to reclassify groundwaters of the state. This has placed the burden of demonstrating the existence of Class IIB conditions onto industry. In the meantime, there is one alternative approach to obtain relief from compliance with Class IIA standards for cleanup of groundwater. This is to petition a reclassification of regional groundwater, a very lengthy, cumbersome process. Reclassification is only considered for large, regional areas. As a result, localized groundwater pollution areas will probably not be candidates for groundwater reclassification. Unfortunately, many of the more remote railroad yards are located adjacent to residential or small urban areas. If the railroad yard and associated industrial complex is not extensive, Class IIA groundwater standards would probably apply to all sites with groundwater pollution. It is questionable whether such conservative standards should apply if there is evidence of historical industrial groundwater pollution and groundwater is not currently used for potable water supplies.

Another consideration when establishing cleanup criteria are the potential matrix interferences of the railyard roadbed construction and fill materials on the currently accepted analytical techniques used to determine closure.   As noted previously, material to be remediated at railroad yards can be comprised of a mixture of native soil, ash, cinders, coal dust, crushed stone and other imported materials.

Remediation progress of diesel fuel contaminated soils is measured using mandated analytical methods.[3]   Infrared (IR) spectroscopy and gas chromatography (GC) measurements of TPH are the current analytical methods of choice for the assessment /screening of diesel fuel contaminated soils, as well as for determining site closure. The TPH-IR (EPA 418.1 modified for soils) is a solvent extraction method. The total mass of hydrocarbon dissolved in the solvent is quantitated by comparing the infrared absorption of the extraction liquid against that of a defined hydrocarbon mixture. Nyer and Skladany[4] have pointed out the following drawbacks concerning the TPH-IR method:

- Volatile compounds are usually lost in the extraction procedure.

- Samples are quantitated against a known hydrocarbon mixture and not the specific petroleum product spilled at the site. All hydrocarbons do not respond equally to infrared analysis, and comparison of the unknown to the standard mixture may result in artificially high or low hydrocarbon concentrations.

- All materials soluble in the solvent will be extracted. These materials may create positive or negative interferences with the hydrocarbon quantitation.

The TPH-IR method is useful as a screening step for determining the presence of hydrocarbon contamination.   However, it cannot identify or quantify   specific compounds present. Its use is not recommended for cleanup standards or certifying that remediation is complete.[4]

Gas chromatography methods for petroleum hydrocarbon analysis include EPA modified method 8015,[5,6] in which method petroleum hydrocarbons are measured by flame ionization detection (FID).   The hydrocarbons that elute during the analysis span the range of approximately C6 to C30. In the evaluation of types of hydrocarbons present in a sample, the chromatogram produced by the sample is compared to a "standard" chromatogram. These standards routinely include gasoline (light fraction), diesel fuel (medium fraction) and motor oil (heavy fraction). Ideally, if a sample of the spilled product were available, it would be used as a standard. The hydrocarbons present in gasoline are comprised of C4 to C10 compounds; those in diesel fuel from approximately C9 to C21; and those in motor oil from approximately C21 to C29. A calibration curve is generated for each standard. From the calibration curve, the concentration and corresponding peak areas of the standards are known. The unknown sample concentration is calculated from the ratio of the total peak area of the sample versus that of the standard. This procedure does not permit quantitation of individual hydrocarbons that may be present, but instead characterizes the contaminant by the "fingerprint" produced by the peak patterns of the hydrocarbons which have eluted and are present in the chromatogram produced by the sample.[7] Assessment of remediation progress at railroad yards by these methods is also difficult due to matrix interferences. Troy and Jerger[8] describe the analytical difficulties encountered when performing bioremediation of diesel fuel contaminated soils at a former railroad fueling yard. Biological landfarming techniques were used to remediate approximately 3,500 cubic yards of a fill comprised of ash, cinders, and crushed stone mixed with soil and clay. Remediation was required as a result of diesel fuel spillage over a number of years. Coal and shale present at the site contained hydrocarbon materials that caused a positive detection by the EPA modified 8015 method. Tripp *et al.*[9] have indicated that unburned coal can be a significant source of environmental hydrocarbon levels.

The foregoing discussion has identified some of problems that must be addressed in order to successfully remediate diesel fuel-contaminated soils in railroad yards. These problems include the following:

- difficulty of establishing background conditions at railroad sites containing heterogeneous fill and railroad bed materials, and in areas with a long history of fossil fuel combustion;

- soil cleanup standards that are often orders of magnitude more conservative than the risks accepted for cleanups at CERCLA and RCRA sites;

- failure of regulatory agencies to identify, classify and develop water quality criteria for degraded groundwaters that have been subject to industrial hazardous waste spills over several decades;

- lack of a process by which to determine compliance with soil cleanup standards;

- inaccuracy of analytical methods used to document the quality of cleanups;

- failure to recognize the importance of natural remediation processes such as sorption and degradation within aquifers;

- inappropriateness of analytical methods such as EPA method 418.1 to accurately determine TPH levels in soil;

- positive matrix interferences with analytical methods due to the presence of heterogeneous materials commonly found in railroad yards such as coal and shale residues;

Fortunately, some northeastern states, such as New Jersey, are adopting regulations which address at least some of these problems. More recognition is required by regulatory agencies of the special problems associated with soil cleanups at railroad yards. If this occurs, regulations can be adopted that truly encourage site cleanups at railroad yards and other industrial areas. Because of the heterogeneous nature of both fill and trackbed materials at railroad yards, background conditions are difficult to identify.

REFERENCES

1. Hay, W. W. *Railroad Engineering. Volume One.* John Wiley & Sons, Inc. New York, New York. 1953.

2. Millner, G.C., A.C. Nye and R.C. James. Human health based soil cleanup guidelines for diesel fuel No. 2. In *Contaminated Soil: Diesel Fuel Contamination.* P. T. Kostecki, E. J. Calabrese, editors. Lewis Publishers, Inc. Chelsea, Michigan. pp.165-216. 1992.

3. Oliver, T. and P. Kostecki. State-by-state summary of cleanup standards. *Soils.* December 1992.

4. Nyer, E.K. and G.J. Skladany. Relating the physical and chemical properties of petroleum hydrocarbon to soil and aquifer remediation. *Ground Water Monitoring Review.* Winter:54-60. 1989.

5. U.S. Environmental Protection Agency. *Test Methods for Evaluating Solid Waste, Physical/ Chemical Methods.* 3rd Edition, SW-846. Washington, D.C. Office of Solid Waste and Emergency Response. 1986.

6. U.S. Environmental Protection Agency. *Test Methods for Evaluating Solid Waste, Physical/ Chemical Methods. 3rd Edition, SW-846. Update 1.* Washington, D.C. Office of Solid Waste and Emergency Response. 1989.

7. Senn, R.B. and M.S. Johnson. Interpretation of gas chromatographic data as a tool in subsurface hydrocarbon investigations. In *Proceedings of the Conference and Exposition, Petroleum Hydrocarbons and Organic Chemicals in Ground Water - Prevention, Detection, and Restoration.* Houston, Texas, November 13-15, 1985. American Petroleum Institute, Washington, D.C. National Water Well Association. pp.331-357. 1985.

8. Troy, M.A. and D.E. Jerger. 1993. Matrix effects on the analytical techniques used to monitor the full-scale biological land treatment of diesel fuel-contaminated soils. Presented at In Situ and On-Site Bioreclamation: The Second International Symposium. San Diego, CA. April 4 - 8, 1993.

9. Tripp, B.W., J.W. Farrington and J.M. Teal. Unburned coal as a source of hydrocarbons in surface sediments. *Marine Pollution Bulletin.* 12:4:122-126. 1981.

CHAPTER 5

Ex Situ Remediation of Diesel Contaminated Railroad Sand by Soil Washing

J. Baker and J. Todd Stanford
Alton Geoscience — Irvine, California

James J. J. Clark
University of California — Los Angeles, California

INTRODUCTION

Soil washing is a rapid, cost-effective method for the remediation of soil contaminated with diesel range fuel. The largest fraction of measurable contaminants, those in the interstitial phase, are rapidly removed by soil washing. The remaining contaminants, primarily the adsorbed phase, require considerably more time to remove and constitute a small fraction of the actual contaminants present.

MATERIALS AND METHODS

Railroad sand was obtained from a railroad site in northern California. The sand, contaminated with diesel fuel, had an average oil and grease concentration of 42,000 parts per million (ppm).

The soil washing apparatus used in the research, shown in Figure 5-1, consists of a galvanized steel trough, an Archimedes screw with changing pitch flights, a motor, a fluid pumping system, and a separator. The screw has flights with alternating pitch to increase the mechanical action in the washing process. The rotation speed of the screw was held constant at 60 revolutions per minute.

As the screw brought soil up the trough, a flow of water/surfactant from jets positioned near the top of the screw would restrict the soil from proceeding upward and would force the soil to move back down the trough. Effluent from the washing process flowed over a weir positioned midway up the auger. After a complete wash cycle (one wash cycle equaled five gallons of water/surfactant mixture), the jets were turned off and the washed soil was allowed to rise up the screw and a sample was collected for analysis. All soils were analyzed for oil and grease by EPA method 413.2.

The soil washing apparatus was engineered such that the angle of the screw, and the vertical position of the weir could be varied. Other variables included in the research were the type of surfactant used, the concentration of the water/surfactant mixture, and the time of washing.

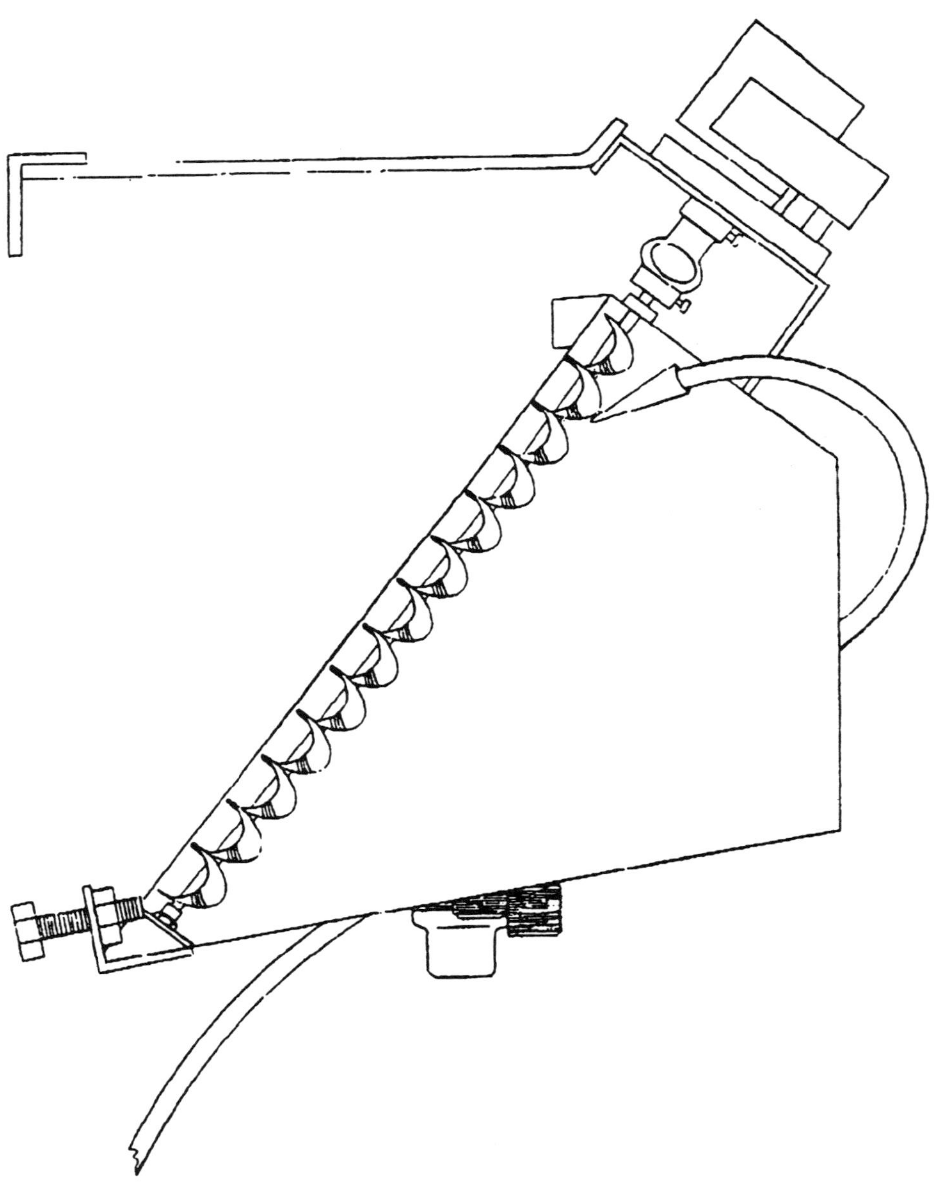

Figure 5-1 Engineering diagram of soil washing apparatus.

Several types of commercially available surfactants were obtained and tested for their ability to extract the diesel fuel from soil. These surfactants ranged from industrial grade degreasing soaps to laundry detergents to dishwashing soaps.

RESULTS

The detergents tested indicated that those surfactants having a high sodium silicate concentration performed best. Sodium silicate increases the pH of the solution and is compatible with the matrix (sand). Higher pH solutions extracted the hydrocarbons more readily. Surfactants high in inorganic components performed better because of the hard surface base of the soil matrix. The surfactant which performed best was a generic dry automatic dishwashing soap (Figure 5-2). This soap, although low in surfactant concentration and high in filler material, was used for our testing of the soil washing apparatus.

Changing the angle of the screw in the soil washing apparatus had little effect on the ability of the system to wash soil. The screw was tested initially at 45 degrees. The angle of the screw was then changed plus and minus ten degrees from the original setting and tests were run on these two angles (Table 5-1).

Like the screw angle, varying the height of the water/surfactant in the trough played a minor role in the washing of the soil. The soil washing apparatus tested had weirs which allowed sampling of contaminated effluent in three different vertical locations. Samples from the three levels did not vary greatly indicating a homogeneous mixing of the contaminant effluent.

The concentration of the surfactant in the water had a large impact on the ability of the soil washing apparatus to extract the hydrocarbons from the soil matrix. Increasing the surfactant concentration increased the ability of the water/surfactant mixture to extract the hydrocarbons. As the concentration of surfactant approached the saturation point of water an attenuation of the ability to extract hydrocarbons was observed. The concentration of detergent in water required to reach that attenuation point is likely to be lower for detergents with less filler, and a higher surfactant to weight ratio.

Time produced the greatest variation in the effectiveness of the soil washing. To determine the efficiency of the soil washing apparatus at extracting hydrocarbons over time, samples of the effluent stream were collected at two minute intervals for a one hour time period and analyzed for oil and grease by EPA Method 413.2. The results of this experiment were then plotted (Figure 3) using log/normal scale.

Table 5-1: Effect of Screw Angle on Soil Washing[1]

Screw Angle (Degrees From Horizontal)	Final Soil Hydrocarbon Concentration (ppm)
35 Degrees	230
45 Degrees	170
55 Degrees	370

[1] Washing performed for 40 minutes using same concentration of detergent.

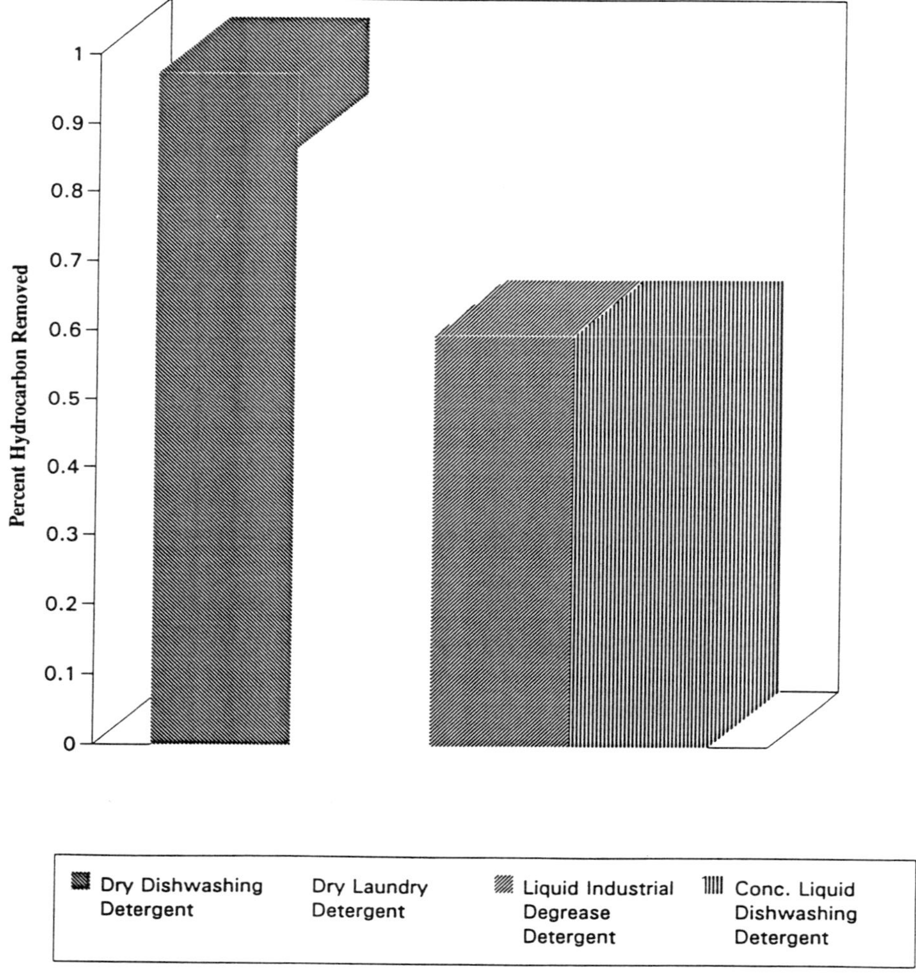

Figure 5-2 Effectiveness of various detergents in soil washing.

Figure 5-3 shows a rapid decrease in sampled hydrocarbon concentration for the first twenty minutes of washing with a slow decrease thereafter. This inflection point at 20 minutes indicates a significant decrease in the effectiveness of hydrocarbon recovery, but more importantly indicates the point at which the interstitial phase hydrocarbons have been removed form the soil matrix. Additional washing, 30 to 40 minutes more, will concentrate on removing the adsorbed phase hydrocarbons which constitute a small fraction of the contaminants. In twenty minutes of soil washing the soil was taken from a contamination concentration of 42,000 ppm to 1,000 ppm, representing a 97.6% reduction. To reach a level below 100 ppm , an additional 0.4% reduction, the soil required one hour of washing time.

DISCUSSION

The predominant phase of diesel in the tested soil matrix was interstitial (approximately 98%). This is evident by the high oil and grease concentration measured. The adsorbed phase hydrocarbons constitute a small portion of the contaminant concentration (approximately 1-2%). A third phase of contaminants present is the vapor phase, which is negligible at ambient temperatures due to the vapor pressure of diesel fuel.

The soil washing apparatus tested had a difficult time washing the adsorbed phase hydrocarbons because of the physical bonding of the adsorbed phase hydrocarbons to the surface of the sand particles. Removal of the adsorbed phase requires a breaking of the bonds and introduction of a matrix with greater affinity for the hydrocarbons than the sand matrix. Coupling soil washing with thermal desorption or freon washing is likely to remove the adsorbed phase but requires added expense and an increase in hazardous waste disposal.

The interstitial phase, the largest fraction of contaminants present, is readily removed by soil washing. Discharged contaminant effluent may be treated by accepted practices (e.g., skimming to remove free product and carbon filtration to remove dissolved phase hydrocarbons prior to discharge). The residual adsorbed phase contaminants may now be dealt with in a number of ways including further remediation (e.g., bioremediation), disposal (e.g., dumping or chemical/physical fixation) or administratively (e.g., risk assessment). Soil washing should be viewed as a first step in the remediation of diesel from sandy soil.

ACKNOWLEDGEMENTS

The authors wish to express their thanks to Christine Hiles for her assistance in the experimentation and manuscript preparation and to Joe Quiros for his aid in the design and manufacture of the soil washing apparatus.

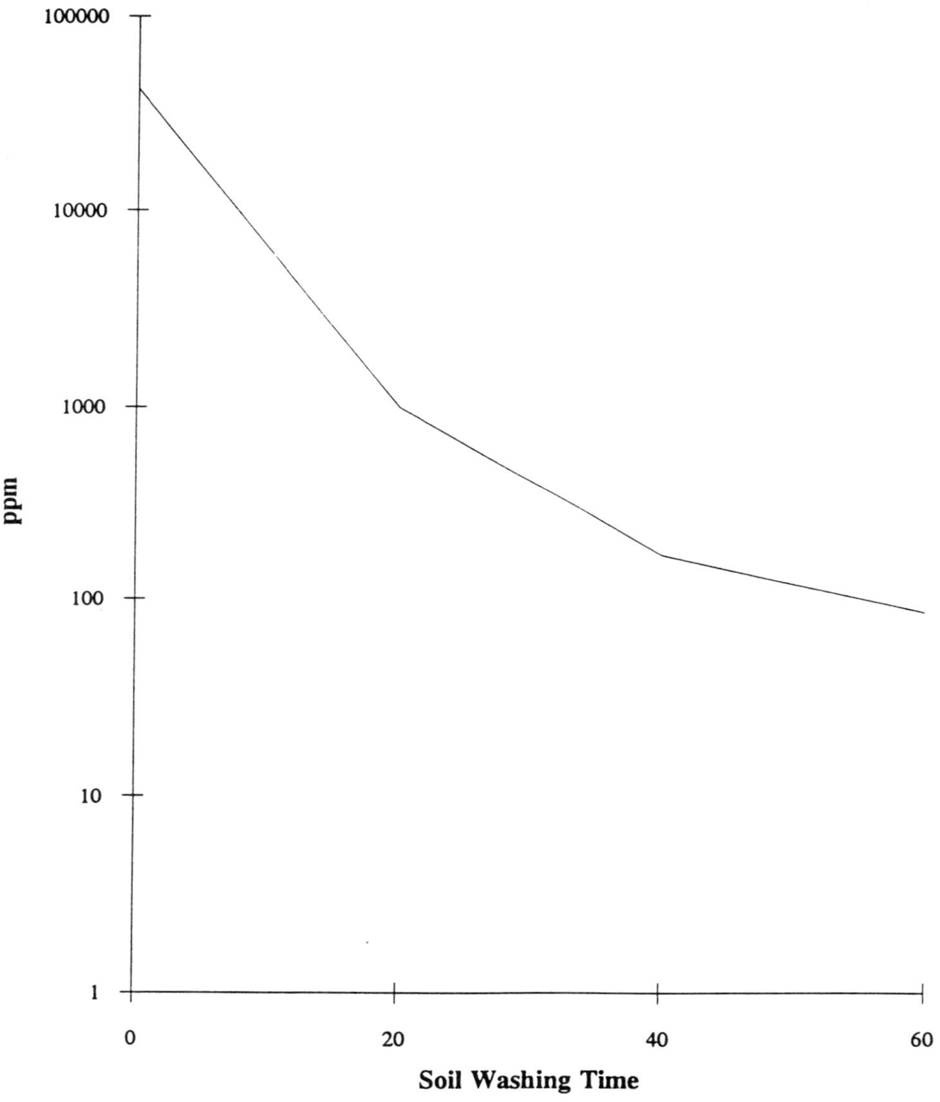

Figure 5-3 Soil hydrocarbon concentration during soil washing.

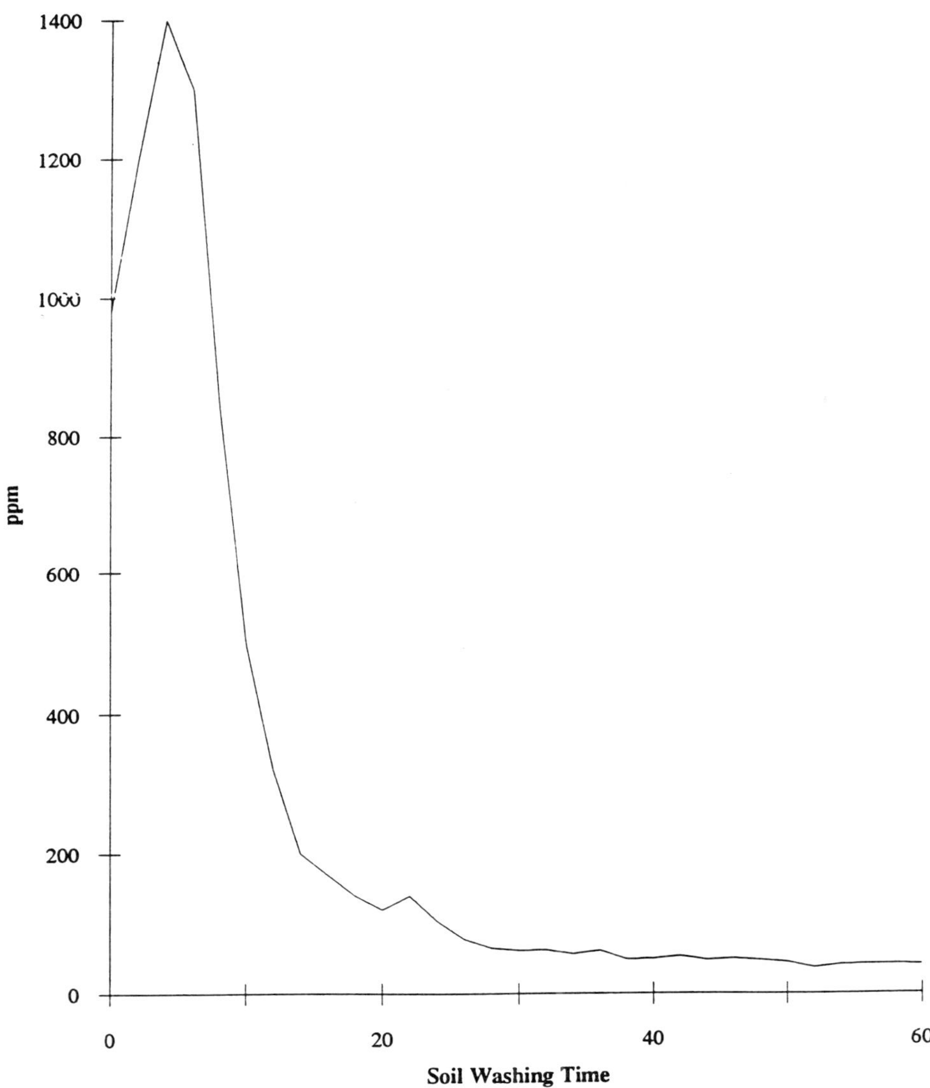

Figure 5-4 Effluent hydrocarbon concentration during soil washing.

Questions and Answers:

Q. What's the cost per ton on your operation?

A. Good question and I can't even give you an answer. All we were evaluating in this particular instance was the application of the soil, and the relative effectiveness of surfactants. We didn't do a cost analysis.

Q. I think you've successfully shown you can remove the hydrocarbons from the soil. My concern is the effluent that's created. You now have a wastewater that's high in oil and grease and high in detergents and ripe for emulsification. Can you shed any light on your success in treating that?

A. Yes and that's a very important point. In essence this treatment alternative really only transfers the contamination from one matrix to another. What is worthy of noting is that we can take this effluent, this surfactant concentration, and recycle this water repeatedly so that we're not reintroducing new, fresh, uncontaminated water into the system repeatedly. Also, through other treatment mechanisms such as, first off, oil water separation and then going on perhaps to a carbon absorption system, there are other approaches. Or if you wanted to take it to the next realm, bioremediation of the contaminated water in a bioreactor vessel. Something along those lines. But it's an excellent question.

Q. Well I had the same question, but did you experience any emulsion problems? Did you look at the treatment of the excess water of the wastewater stream?

A. In a nutshell, no. What we did was run the water through an oil water separator and then just use carbon absorption. We didn't observe any problems with not being able to polish that effluent water stream.

Q. I had another question about pH and the cation exchange capacity of your soil. Did you make any attempt to correlate the cation exchange capacity of your soil with the ionic charges of your different surfactants?

A. No, we did not. Important point. Lesson to be learned for the next round of analysis on this particular instance.

CHAPTER 6

A Field Evaluation of Remediating and Recycling Railroad Ballast Using a Modified Soil Washing Technique

L. Joy Lozano
Tesoro Environmental Products Company

M. Neal Guentzel, Ph.D.
The University of Texas at San Antonio

INTRODUCTION

Soil washing is a developing technology for removing petroleum hydrocarbon contamination from soil and rock particles. The U.S. Environmental Protection Agency (EPA) has identified soil washing as an alternative treatment method to reduce the quantity and impact of pollutants in soil[1]. By definition, soil washing uses mechanical and/or chemical means to disperse hydrocarbon contamination from soil and rocks and to isolate the contaminants into as little volume as possible[2]. This technology is becoming more widely used in the oil industry as a recycling and waste minimization tool and has application in many other industries including track remediation in the railroad industry.

Tesoro Environmental Products Company (TEPCO), a subsidiary of Tesoro Petroleum Company, markets several remediation products. PES-31® and PES Nutrient Supplement are for general bioremediation of contaminated soils and waters. PES-41® is formulated to lift hydrocarbons off of porous and nonporous hard surfaces, followed by biodegradation. PES-51®, a biological cleaner for lifting oil from rocks and bulkheads following ocean spills, allows oil to be skimmed or absorbed by booms or pads. The Tesoro Refinery in Kenai, Alaska has used soil washing with PES-51® and PES-41® as a waste minimization and recycling tool for contaminated rocks in combination with PES-31® for bioremediation of the soil and water phases. This study describes an evaluation of soil washing technology using TEPCO's PES-51® for cleaning heavily contaminated railroad ballast. Results indicated that hydrocarbon contamination of the ballast rocks was reduced by 86.5% to 96.4% in the five test runs with an average reduction of 92.6%. This study suggests that the use of appropriate ballast washing technology in combination with PES-51® can reduce or eliminate the need for costly disposal and replacment of old ballast material, encouraging reuse and recycling.

MATERIALS AND METHODS

Test Site and Test Material

The site chosen for this study was the Cooper Equipment Company, San Antonio, Texas. Eight tons, approximately 6.4 cubic yards, of heavily contaminated railroad ballast material was obtained from a Union Pacific railroad yard in Dennison, Texas two days prior to initiation of the tests and was delivered to the test site one day prior to initiation of the tests. The contaminated ballast, which is shown in Figure 6-1, was dumped onto a 20 ml high density polyethylene (HDPE) lined pad and moved with a front-end loader. The ballast material, shown in close-up in Figure 6-2, contained a high content of fines, rocks from the size of gravel to about 4" in diameter, balls of tar-like material, railroad spikes and other industry-related debris.

Source of Other Materials Used for the Study

Two 55-gallon drums of PES-51® were obtained from Tesoro Environmental Products Company. A Royer Shredder-Separator Model 300, a Coyote Model C18 front-end loader, and a Furukawa Model 12D front-end loader were obtained from Cooper Equipment Company. Three 6' x 2' x 2', 180 gallon capacity, galvanized steel tanks were obtained from Builders Square. A 4' x 1.5' x 2.5' steel mesh (1/4" X 3/8" diamond pattern) basket, with an opening gate in the front for dumping the cleaned rocks and designed to fit the tanks and the Furukawa front-end loader as shown in Figure 6-3, was fabricated by Cooper Equipment Company. San Antonio ambient tap water, from the Edwards Aquifer, was used in the washing process.

Sampling and Analysis for Hydrocarbon Contamination

Samples were collected in 1,000 ml jars with Teflon®-lined metal tops. Three multi-spot, composite samples of the original material were obtained by using a small metal scoop to collect random samples from all over the pile until the container was filled; one initially, one after about 40% of the pile was treated, and the last after about 80% of the pile had been removed for the first four test runs. A composite sample of treated ballast was taken in a similar fashion from each of the five replicate test runs, after completion of the respective test run. Additionally, two composite samples of cleaned ballast were taken by random sampling from each of the five test piles at the end of the study. Samples of the water from the rinse baths after use also were taken.

Samples were packed in ice in coolers and shipped to an independent analytical laboratory. The chemical analytical test method for hydrocarbon contamination used in this study was TRPHC (Petroleum Hydrocarbons, Total Recoverable, Spectrophotometric, Infrared) by EPA Method 418.1 (Texas Natural Resource Conservation Commission).

Ballast Washing Procedure

Figure 6-4 is a schematic overview of the ballast washing procedure. The eight tons of heavily contaminated ballast were processed in five batches. Separate batches were loaded into the Royer hopper using the Coyote front-end loader. The variable sweep of the Royer was adjusted to the smallest setting because of the small size of the ballast material and the shredding conveyor belt was reversed periodically to maintain optimal flow and separation. The Royer Shredder-Separator separated the contaminated ballast material into two phases, contaminated rocks and soil debris (Figure 6-5).

Figure 6-1. Initial pile of contaminated ballast material.

Figure 6-2. Close-up of contaminated ballast material. A nickel is included for perspective.

Figure 6-3. Mesh basket supported by front-end loader.

Schematic of Railroad Ballast Washing Procedure

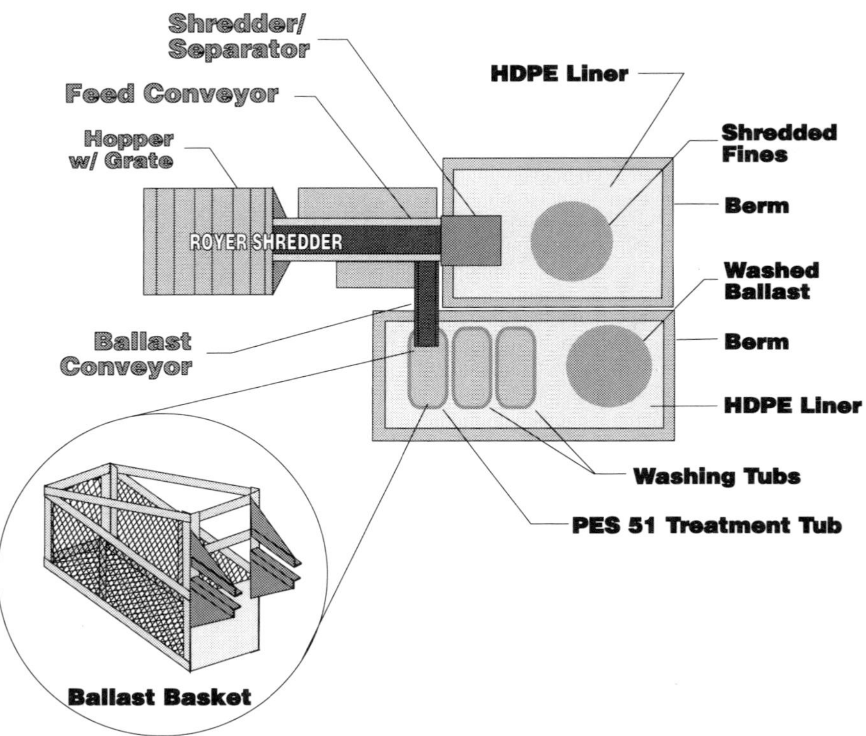

Figure 6-4. Schematic of ballast washing procedure.

The contaminated ballast rocks were discharged into a bath of a PES-51® sufficient to cover the top layer of rocks. The maximum processing rate of the Model 300 is 75 cubic yards/hour, therefore, each batch of ballast was processed by the Royer in approximately 5 minutes. The rocks were agitated by moving the mesh basket up and down approximately ten times. The basket was then transferred to the water bath, agitated as before, and allowed to sit approximately 15 minutes as the top layer of oil was skimmed off with a pump. The first and second batches were processed with a single wash. The last three batches were processed the following day with two washes in separate tanks. In every case, the wash tanks were drained and refilled between batches with fresh water. The PES-51® tank was drained after the processing of the first two batches of ballast on the first day. The tank was refilled with fresh PES-51® on the second day for processing the last three batches of ballast material without changes of fresh PES-51® between batches. The cleaned ballast material was dumped into separate lined areas for each batch processed.

RESULTS

The PES-51® treatment markedly reduced the level of contaminating hydrocarbon on the ballast rocks as shown in Figure 6-6. The average TRPHC level for the original three composite samples of contaminated ballast was 30,491 ppm. The levels measured for the PES-51® treated ballast range from a high of 4,106 ppm (first batch treated) to 1,090 ppm (last batch treated). Figure 6-7 shows the data plotted as the percentage of the original TRPHC measured for the contaminated ballast. Hydrocarbon contamination of the ballast rocks was reduced by 86.5% to 96.4% in the five test runs with an average reduction of 92.6%.

The levels of contamination measured on the two composites (made from the two different rinsing techniques) of the five treatment runs showed a similar level of reduction of hydrocarbon contamination (Figure 6-8). The TRPHC of the original material (30,491 ppm) was reduced to 2,493 ppm (91.8% reduction) in the first composite and 1,513 ppm (95.0% reduction) in the second composite.

An "oil slick" of hydrocarbon contaminants was produced on the surface of the rinse waters following the PES-51® treatments. The oil slick was apparent almost immediately upon placing the basket of rocks into the wash rinse tanks. Black globules rose to the surface and soon formed a thick "slick" which could be skimmed off of the water surface (Figure 6-9). The extent of the recoverable oil for three of the rinses is shown in Figure 10. The recoverable oil, measured as a percentage and ppm of the total oil and water volume, varied from approximately 6,000 ppm to over 200,000 ppm.

Figure 6-5. Separation of contaminated ballast material into contaminated rocks and soil debris by the Royer Shredder-Separator.

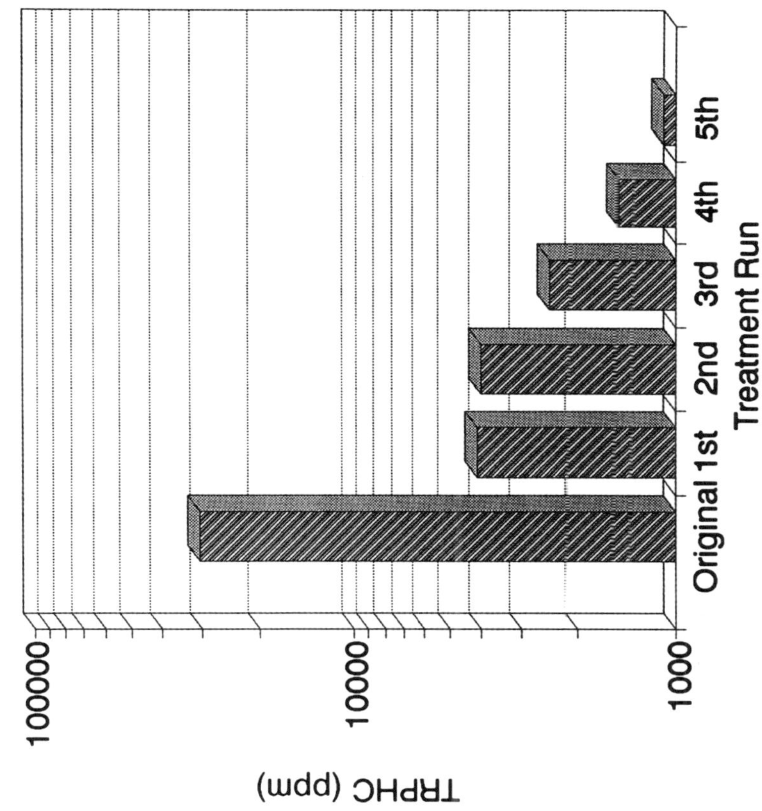

Figure 6-6.   Reduction of total recoverable petroleum hydrocarbons (TRPHC, 418.1) in contaminated ballast by treatment with PES-51®; five separate treatment runs.

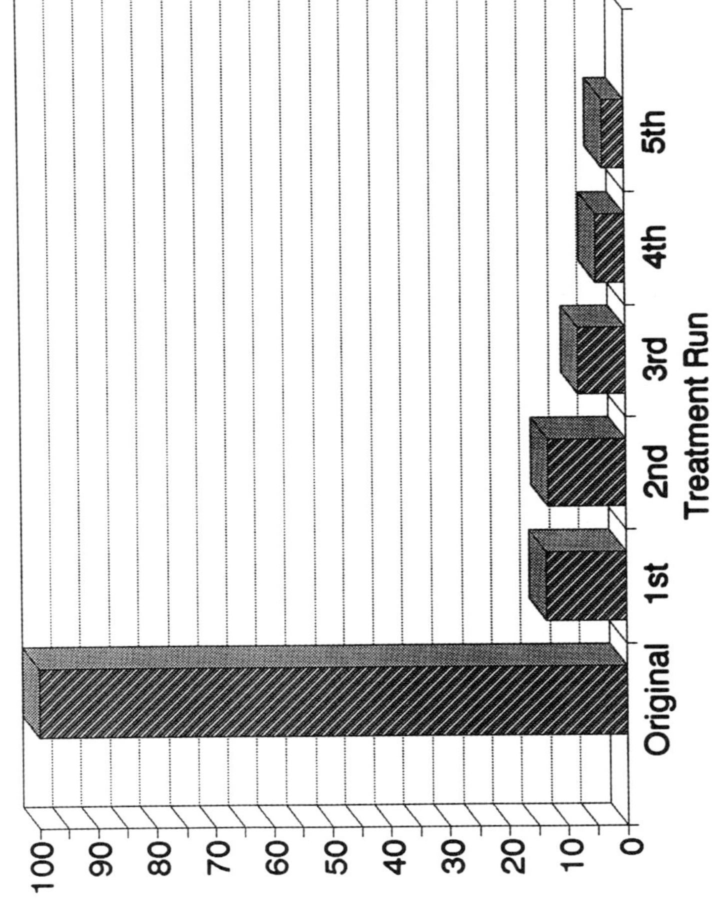

Figure 6-7. Percentage of original total recoverable petroleum hydrocarbons (TRPHC, 418.1) in contaminated ballast after treatment with PES-51®; five separate treatment runs.

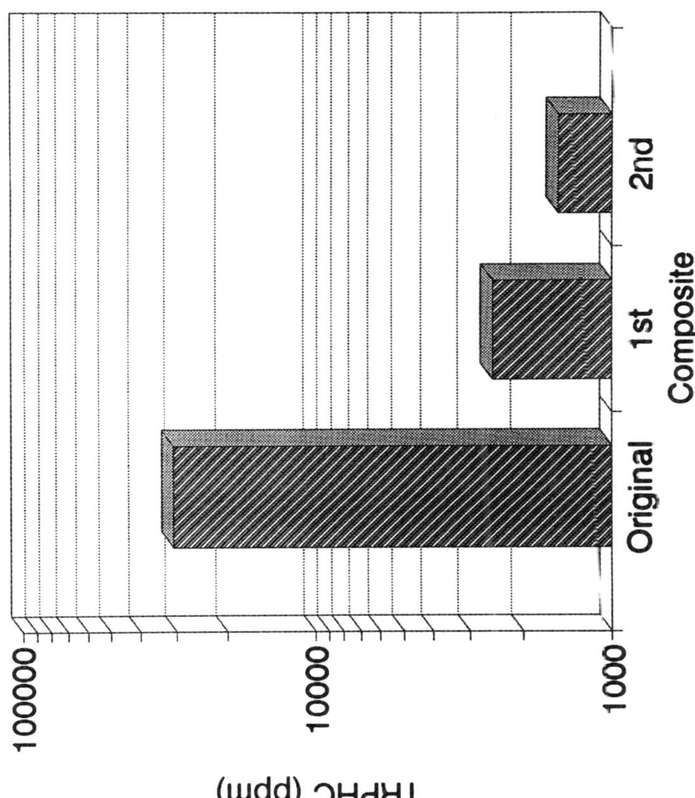

Figure 6-8. Levels of contamination as total recoverable petroleum hydrocarbons (TRPHC, 418.1) in two compositions of five treatment runs with PES-51®.

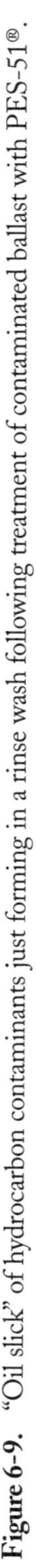

Figure 6-9. "Oil slick" of hydrocarbon contaminants just forming in a rinse wash following treatment of contaminated ballast with PES-51®.

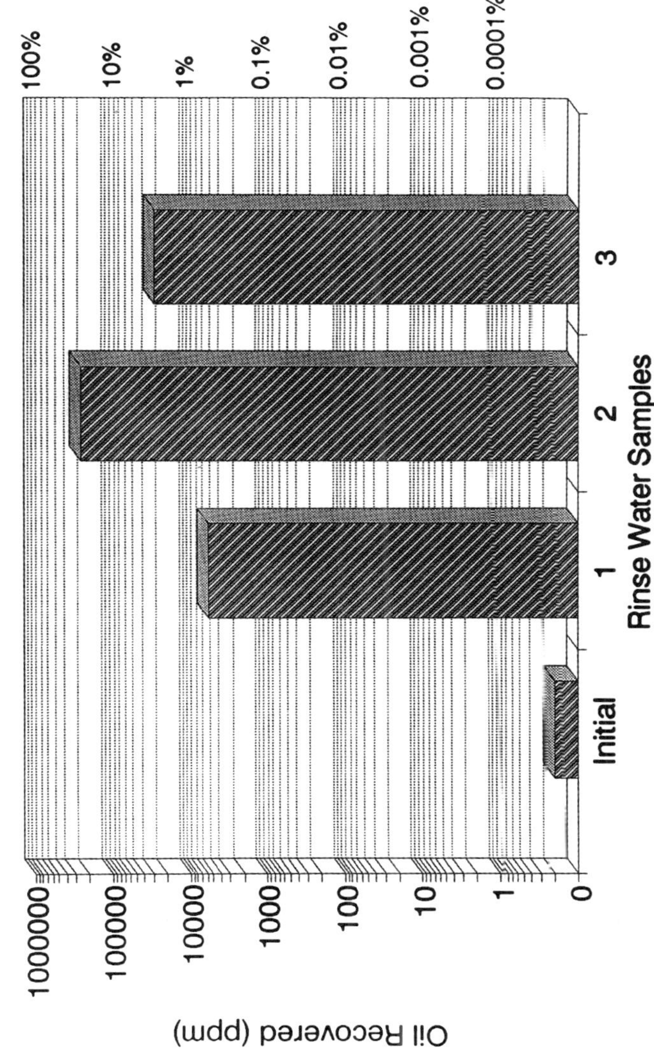

Figure 6-10.   Extent of recoverable oil, measured in percentage and ppm of the total oil and water volume.

DISCUSSION

Hydrocarbon contamination of railroad ballast is an important concern for the railroad industry. The results of this chapter demonstrate that a washing technology using TEPCO's PES-51®, a natural biodispersant/biosurfactant, markedly reduced the level of hydrocarbon contamination of ballast by up to 96.4%. It also is apparent from Figures 6-6 thru 6-7 that we became more proficient in cleaning the ballast material with each batch processed. Addition of the second wash for the last three batches processed had a major effect. Longer exposure to PES-51® with better agitation of the rocks, and a high pressure spray wash are simple modifications that may produce even more impressive results. The "batch process" used in this study can be upsized in a number of ways. For example, the Royer Shredder-Separator Model 300 used in this study operates at a maximum processing rate of 75 cubic yards/hour, whereas the Model 401 operates at a maximum processing rate of 200 cubic yards/hour. A continuous processing track machine also could be developed.

Recycling of used PES-51® in processing the last three batches did not decrease its effectiveness and the extent that this recycling is possible should be determined to make the process more cost effective. Additionally, rinse water with the oily layer removed through skimming could be reused for primary washing with the oil itself being recycled/recoved. An important point concerning the present study was that we chose to use ballast that approached a "worst case scenario;" that is, it consisted primarily of smaller rocks with a very high content of debris, fines and asphaltic tar and tar balls. A more conventional ballast may be much easier to clean.

Dove, et al.[2] in their study of soil washing to remove petroleum hydrocarbon contamination from soils, observed that a variety of factors may influence the efficacy of the washing process for soils. These factors which influenced hydrocarbon retention included particle size, humic acids, metal oxides, geological origin of the soil particles, and clay content.

With appropriate modification of the technology described in this chapter it may be practical and cost effective to clean and recycle hydrocarbon contaminated ballast material along the tracks or even in place, eliminating the need for costly replacement and current environmentally unfriendly disposal alternatives such as dig and haul; stockpile; or incineration.

REFERENCES

1.	Daley, P. S. 1989, Cleaning up sites with on-site process plants, *Environ. Sci. Technol.* <u>23</u>:912, 1989.

2.	Dove, D., A. Bhandari, and J. Novak. 1992. Soil washing: Practical considerations and pitfalls. *Remediation*, 55, Winter 1992/93.

ACKNOWLEDGEMENTS

The authors want to thank George Cooper of the Cooper Equipment Company, San Antonio, Texas, for his extensive help in this study. This work was supported in part by a Sponsored Research Agreement between TEPCO and The University of Texas at San Antonio (Dr. M. N. Guentzel).

Questions and Answers:

Q. Do you know how the analytical lab handles the rocks?

A. We have our own analytical capabilities, but we sent this to an independent laboratory. A couple of things, first the TPH that we ran reads up to C-26. That's normal. Anywhere from C-10 all the way up to C-26. My understanding is they put this through an extract they do not pulverize the rock they simply wash it with the freon extract trying to pull everything off and in our case this was not a porous material so we feel like it was extracted completely off the surface.

Q. Did you run the mixture through the separator, pull the fines out, and run your test on just the ballast rocks?

A. On the ballast rocks, that's exactly right.

Q. Was there any attempt to look at the fines and address those?

A. No, but this could potentially be done if you had a ballast washing machine that was moving along the tracks and was pulling up the ballast, treating it and putting it back down. Again these are short term treatments required for removal of the hydrocarbon contaminant from the rock. Those fines could then be treated, for example by, a general bioremediation product such as TS31.

Q. Why weren't the fines included with the ballast in the washing test?

A. Because we were interested in looking at the ballast itself and separating out the fines from the ballast material.

Q. I had a couple of questions. Maybe I missed the part about your *ex situ* washing. What was the contact time of your ballast rocks with your PES solution?

A. To measure the contact time in the PES solution, we took the rocks which fell into the basket. The processing rate for this machine is 75 cubic yards per hour and we're dealing with a much smaller volume in each batch than the maximum processing rate would handle in an hour, so it only took 5 minutes for the rocks in the ballast material to end up being deposited into the basket, so potentially there was as much as five minutes exposure there. Then the rocks were agitated by moving them up and down ten times. That was probably another five minutes. So maximally the rocks were exposed to the PES 51 for about ten minutes and perhaps somewhat less.

Q. What was the contact time of your ballast rocks during the *in situ* treatment when you were spraying the railroad bed?

A. It was two minutes.

Q. Why wouldn't you have run a combined product where you have the microbes as well as the cleaning agent, and just spray it on the ballast and let it take its place and then test the ballast underneath to see if it degraded the oil?

A. Well, what we're hoping for is the development of a machine that would rapidly clean the ballast material as the machine would perhaps move down the track or on the side of the track and redeposit the ballast material as it's pulled up. Now in terms of using 31 to treat the ballast material, this would be a long-term process when you're looking at biological degradation versus chemical, mechanical removal. So using the 31 for this process alone would not have been a time effective type of operation.

Q. But it would have had the effect of removing it from site while the microbial action was taking place?

A. The wastewater after the oil was skimmed off didn't have a high concentration of TPH. That certainly could be treated and recycled and used for washing because we can pull off the oil and send it back to a refinery and use the water for rewashings and initial rinses and then that water, the wastewater, could be treated with 31. The concentrations are not so great; it should take a relatively short period to treat the wastewater.

Q. I'd like to make a comment on that last question. It's very difficult to bioremediate ballast material. There's really no substrate to support microbial growth. Only at a depth of maybe 2 or 3 feet are you going to be able to get enough fines or soil type materials to support microbial growth. We've done this at a number of locations where we do a two-pronged approach, where we use the PES 41 or 51, to treat the topical aspects of the track, come back and reapply the PES 31. Well, first after we apply the 51, we remove the oil, vacuum it up, it's off in the trenches. After that's been cleaned and allowed to dry, we come back with the PES 31 application with the intent of letting that percolate down and biodegrade the contaminants at depth.

A. I'd like to expand on Joy's comments, an issue she's raised is very important. If you look at the idea of bioremediating material that's like the heavily contaminated ballast material that we used here that contains large concentrations of tar balls and other types of material. Microbes aren't going to swim inside of those tar balls, they're only going to degrade on the surface and that's a limiting factor. The material that the microbes are degrading is hydrophobic. Microbes, by and large, are hydrophilic and they're going to degrade at that hydrophobic/hydrophilic interface, the aqueous interface. If you can create more surface area by breaking that up, then you can enhance the degradation rate. This is how surfactants work to enhance degradation. But just using microorganisms straight on that heavily contaminated ballast, the rate of degradation would be very very slow just by nature of the fact that you have that type of material. The organisms would degrade on the surface. It would be a very very slow process before it was all degraded.

CHAPTER 7

Use of Bio-Slurry in the Construction of a Diesel Recovery Trench

Mick Hardin
The Atchison, Topeka, and Santa Fe Railway Company

Bob Fronczak, Gary Henderson and Jeff Bordelon
Radian Corporation

INTRODUCTION

An active railroad fueling facility and switching yard in central New Mexico is the site of diesel fuel contamination of soil and groundwater. Historic releases from fueling facilities and associated diesel fuel pipelines throughout the yard have resulted in three separate, free-phase diesel fuel plumes on the shallow groundwater table across the site. One of the diesel fuel plumes extends beyond the railyard property boundary. The owner of the yard is currently implementing remedial measures to remove the free-phase diesel fuel from the shallow water table aquifer. This chapter briefly describes the environmental setting of the yard and presents a summary of remedial measures currently being implemented at the yard. An emphasis is placed on the construction of an interceptor/collector trench using biodegradable polymer slurry to stabilize the walls of the excavation.

GEOGRAPHIC AND GEOLOGIC SETTING

The yard is located in the Rio Grande Basin approximately 30 miles south of Albuquerque, in Belen, New Mexico. The yard is an active fueling facility and switching yard that handles anywhere from 35 to 50 trains a day. The yard is located at the eastern edge of downtown Belen. Residential areas bound the facility on the east along the length of the yard. On the west side, a commercial district and residential area bound the yard. The yard is approximately one mile from the Rio Grande River.

The yard lies within the 100 year flood plain of the Rio Grande River and is underlain by more than 100 feet of alluvium. The alluvium in the area of the yard consists of unconsolidated sand, silt, gravel, and clay. The alluvium is underlain by sediments of the Tertiary age Santa Fe Group.

The depth to groundwater across the site varies from approximately four to 10 feet below surface. The shallowest groundwater occurs at the south end of the yard where the surface is at a lower elevation than in other portions of the yard.

Figure 7-1.   Diesel Fuel Thickness Contour Map, June 1993.

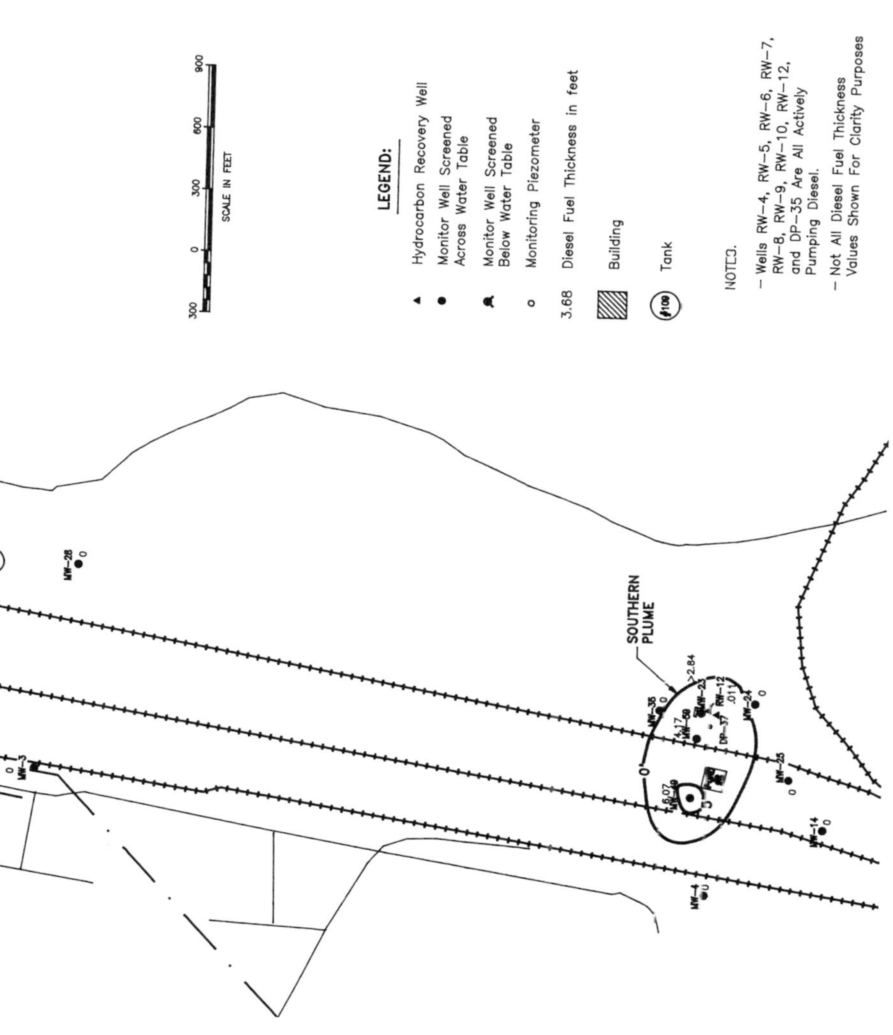

Figure 7-1. (Cont.)

Groundwater flow in the upper alluvium is southerly and generally parallels the Rio Grande River. The gradient is estimated to be approximately 0.001 and the groundwater flow rate estimated to be about 80 to 100 feet per year. The shallow alluvial aquifer has been measured to have a transmissivity of 50,000 gallons per day per foot (gpd/ft) and a storage coefficient of 0.14. Aquifer pumping tests in the upper 10 feet to 15 feet of the aquifer yield values for transmissivity of 20,000 gpd/ft.

OCCURRENCE OF DIESEL FUEL IN THE SUBSURFACE

Three separate free-phase diesel fuel plumes have been identified to date across the site. These plumes are designated as the Northern, Central, and Southern Plumes. Figure 7-1 shows the location and configuration of the three diesel fuel plumes.

Northern Diesel Fuel Plume

This plume is the smallest of the three plumes. The long dimension of the plume, northwest-southeast direction, is approximately 450 feet and the width of the plume is estimated to be approximately 250 feet. The measured diesel fuel thickness in the monitoring wells in this plume are on the order of one foot or less. Recovery testing shows that diesel fuel migrates slowly into the well casing after bailing, suggesting that free product recovery in this area will be slow. Geologic data indicate that the capillary fringe zone is relatively fine-grained in this area, causing the slow recovery after bailing.

Central Diesel Fuel Plume

This plume is the largest plume on the site and is estimated to be approximately 2,000 feet in length (northwest-southeast) and 500 feet wide at the widest point. The maximum measured thickness of diesel fuel in monitoring wells is between 5 and 6 feet. Geologic data indicate that the capillary fringe zone at the eastern margin of the Central Plume is sandy. Diesel recovery tests indicate the potential for high recovery rates in the eastern margin of the Central Plume, because of the sandy soil.

Southern Diesel Fuel Plume

The configuration of the Southern Plume is different from the Northern and Central Plumes. Whereas the latter two plumes are oriented lengthwise in the hydraulically downgradient direction (south to southeast), the Southern Plume appears elongated in the east-west direction and limited in the hydraulically downgradient direction. The configuration of the Southern Plume may be controlled more by the underground pipeline orientation than the groundwater flow direction. The Southern Plume is approximately 650 to 700 feet in an east-west direction and about 500 feet in a north-south direction. The maximum measured thickness of diesel fuel in the Southern Plume monitoring wells is 7 feet. Initial recovery testing in this plume suggest variable recovery rates.

DIESEL FUEL RECOVERY

The initial recovery of diesel fuel at the railyard has been through passive skimming in recovery wells. The recovery well network includes five eight-inch diameter recovery wells installed in the 1,030-foot long hydrocarbon recovery trench (HRT) along the eastern margin of the Central Plume, 3 recovery wells installed in augured boreholes, and an old excavated well from the early 1980s.

Hydrocarbon Recovery Trench

The HRT was installed to facilitate the recovery of free-phase diesel floating on the groundwater along the eastern margin of the Central Plume. The main components of the HRT installation included a recovery trench, hydrocarbon recovery wells, monitoring piezometers, very coarse sand backfill, and geotextile fabric (Figure 7-2). Five hydrocarbon recovery wells (8-inch diameter) and six monitoring piezometers (2-inch diameter) were installed within the trench.

Because of the unstable sandy soils present at the site, the HRT was excavated using a bio-polymer slurry to support the trench walls. The highly viscous bio-polymer slurry biodegrades into sugar and water, leaving no chemical residue in the subsurface. The bio-polymer slurry used for this project was made by mixing bio-degradable guar gum powder with water.

The completed length of the trench was approximately 1030 feet. The HRT was approximately 11 feet deep (4,790-feet MSL) along its entire length except in the vicinity of the recovery wells. The trench was excavated to a depth of 16 feet for a span of 20 feet at each recovery well location. A deeper trench was required at the recovery well locations to allow enough clearance for the hydrocarbon recovery pumps and for the future installation of submersible groundwater pumps.

The HRT was backfilled with a very coarse sand with the following gradation:

- 100% passing the No. 3/8-inch sieve;

- 18% passing the No. 4 sieve;

- 2% passing the No. 8 sieve; and

- 0.8% passing the No. 16 sieve.

The backfill material was used to fill the trench to within two to three feet of the surface. A geotextile filter fabric (70-100 mesh) was placed on top of the sand backfill before filling the remainder of the trench with clean soil.

On the day preceding the commencement of trench excavating, several thousand gallons of bio-polymer slurry were prepared. Powdered guar and water were mixed using a venturi tube type high shear mixer. Approximately 50 pounds of guar were added to every 800 gallons of water. The slurry was mixed for several hours and stored in a 20,000 gallon tank.

The viscosity of the bio-polymer was maintained around 63 seconds (Marsh funnel). A very small amount of acetic acid was added to the mixture in order to adjust the pH for maximum preservation of the slurry. The pH of the slurry varied from 5 to 5.5 during the trench construction.

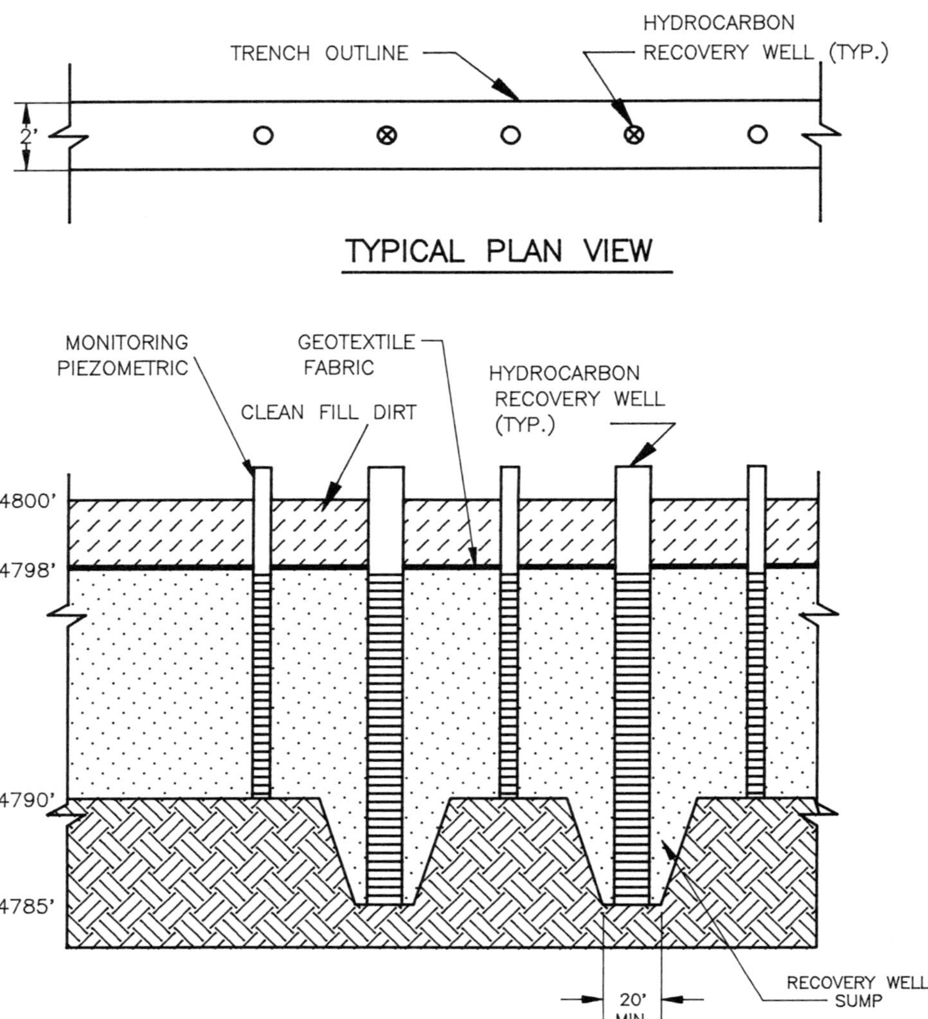

Figure 7-2. Hydrocarbon Recovery Trench Cross-Section.

Excavation of the trench began on 27 January 1993 at the south end of the trench. A track-hoe excavator with a two-foot-wide bucket was used to excavate the trench. The excavator operator constructed the trench in 20-foot segments. In each segment, the top two to three feet of clean soil removed from the excavation were used to build a small berm on both sides of the trench. At that point, slurry was introduced into the trench segment. All soil excavated from the trench which contained slurry was loaded directly into dump trucks for disposal at a landfarm.

The backfill material (very coarse sand) was placed directly into the slurry-filled trench using a front-end loader. Backfilling was conducted as soon as possible in order to minimize the length of time where the trench walls were only supported by the slurry. The trench was backfilled with the coarse sand to within two feet of the ground surface.

The recovery wells and monitoring piezometers were installed in the HRT using a loading hopper. Each well casing was lowered into the hopper and centered in the trench. A hand level was used to check the vertical alignment of the casing. Rubber cords were used to secure the well casing in place. Backfill material was then poured into the hopper and around the well casing using the front-end loader. Enough backfill material was placed around each well so that the hopper could be removed from the trench.

All excavating and backfilling of the HRT was completed by 29 January 1993. In order to reduce the viscosity of the slurry remaining in the trench backfill, a liquid enzyme breaking agent was poured directly into the recovery wells and along the trench itself. Approximately five gallons of breaking agent were used for this procedure. The day after the breaking agent was added, a small submersible pump was used to circulate the slurry from each of the recovery wells to ensure the complete breakdown of the slurry. The discharge from the submersible pump had the appearance and viscosity of groundwater indicating that the slurry had been broken down.

The geotextile filter fabric was installed in the trench after the slurry was broken down. The filter fabric was rolled out on top of the coarse sand backfill material. The clean soil excavated from the trench was placed on top of the filter fabric and compacted to 90% of the maximum dry density.

Drilled Vertical Recovery Wells
Three drilled hydrocarbon recovery wells were installed at the Belen yard in late January 1993. Two wells were installed to recover diesel fuel from the Central Plume and one well was installed in the Southern Diesel Fuel Plume. These wells were located in areas where trenching could not be performed because of yard traffic and structures.

All of the interior recovery wells were constructed of 8-inch diameter stainless steel well casing and screen. A 20-foot section of 0.020-inch slot well screen was installed for each well. A 10-20 mesh silica sand was used for filter pack around the well screen. A one-foot thick bentonite seal was placed above the filter pack. Cement-bentonite grout was used to fill the remainder of the annular space in the well. A hollow-stem auger drill rig was used to install the recovery wells. All five wells were drilled to a depth of 25 feet below the surface using 14-inch O.D. augers (10.25-inch I.D.).

Once the final depth was reached, the well screen and casing were placed inside the augers. The sand filter pack was poured through the augers to a level approximately one foot above the top of the screen slots. The bentonite chips were placed above the filter pack, hydrated with water, and allowed to set up for a minimum of 30 minutes before placement of cement-bentonite grout.

Each well was developed, after allowing the grout to set up for a minimum of one day, using a portable surface centrifugal pump. Approximately 3300 gallons of water were pumped from each well at an average rate of 35 gpm. The development of each well achieved the desired objective of minimal sand production.

Diesel Fuel Recovery Systems

The new recovery wells at the railyard are equipped with product-only skimmer pumps. The pumps used are completely electric and utilize an oleophyllic/hydrophobic screen that allows hydrocarbon, but not water, to pass through. Only diesel fuel is pumped. Diesel fuel recovered from the recovery wells is stored in on-site reclaim tanks and eventually loaded into rail tank cars and recycled.

SUMMARY AND CONCLUSIONS

Initial diesel fuel recovery in this New Mexico railyard has been implemented using a 1,030 foot long hydrocarbon recovery trench with 5 recovery wells and 3 drilled recovery wells. The hydrocarbon recovery trench was installed using a bio-polymer slurry to stabilize the trench walls so that recovery wells could be installed. The excavation and construction of the trench and recovery wells was greatly enhanced by the use of the bio-polymer slurry.

After two months of operation approximately 17,800 gallons of diesel fuel have been recovered from the subsurface at the railyard. The recovery of diesel fuel in the hydrocarbon recovery trench stands at approximately 17,200 gallons showing that the vast majority of diesel fuel is being recovered from the trench rather than the drilled recovery wells.

Questions and Answers:

Q. Have you ever experienced problems with your trench filling up and then exiting around the sides?

A. No we didn't. You mean actually overflowing the trench?

Q. Going around, exiting the very ends of your trench.

A. No we controlled the level in the trench so we didn't have any overflow problems at all.

Q. Is there a constructive use for the product that's recovered?

A. Normally what happens is that it goes into a railcar that's transported back for recycling for reuse. Generally there's a value to that product.

CHAPTER 8

Biological Soil Recycling in a Fixed Facility:
A Railroad's Perspective

Mark R. Murphy, REM
CSX Transportation

James Novitsky, Ph. D.
KEMRON Environmental Services, Inc.

Stephen G. McMahon, P.G.
Integrated Environmental Solutions, Inc.

INTRODUCTION

CSX Transportation (CSXT), as one of the United States' largest integrated multimodal transportation companies, pumps approximately one million gallons of diesel fuel into its locomotive fleet on a daily basis, 365 days a year. In consideration of its current operations, past operating practices, and inappropriate property usage by lessees, CSXT presently faces a tremendous task in remediation of numerous diesel-contaminated sites. The Environmental Remediation Group views this challenge as a tremendous opportunity to develop and apply new remedial technologies.

However, like any other department in the railroad, CSXT's Environmental Remediation Group must compete for scarce resources in order to complete its intended mission. Recognizing that it has a large quantity of diesel contaminated soil, CSXT over the years has evaluated and tried numerous remedial options. In applying its own Remedial Alternative Evaluation System to the process of diesel contaminated soil remediation, two methods are now being fully field evaluated. Both of these methods represent low cost remedial alternatives. This chapter presents an analysis of one of these alternatives, ex-situ biological remediation in a fixed facility.

BACKGROUND

CSX Transportation and its predecessor railroads have operated within Savannah, Georgia and the Chatham County area since the late 1850's. These operations have included full rail service, locomotive repair shops and yards, signal shops, carpenter shops, and other essential facilities. Over the years via consolidation of operations and general downsizing, a number of these facilities have become redundant. During the closing of many of these facilities, CSXT recognized the need to remediate large quantities of petroleum, mainly diesel contaminated soil. Compounding the problem was the inappropriate use of many CSXT properties by lessees. CSXT's problem, by today's estimate, involves over 60,000 tons of contaminated soil in Chatham County alone. System-wide the problem is worse, industry-wide the problem is enormous.

The conventional remediation method of soil incineration would cost CSXT approximately $1.4 million to $1.75 million dollars for this one county. Using an alternative technology, could CSXT remediate the contaminated soil at lower cost and redirect these cost savings to other remedial efforts? A method of remediation that has promise from an economic perspective and which has been extensively evaluated by CSXT is ex-situ biological remediation in a fixed facility. Ex-situ biological remediation offers a low cost, low maintenance, permanent, and environmentally appealing approach to recycle petroleum contaminated soil.

The two most significant factors affecting this avenue of remediation were first, was the state of Georgia receptive to this type of facility, and second, was an adequate facility available? Initial informal discussions were held with the Georgia Environmental Protection Division and a positive interest was expressed by the State. Subsequent technical and regulatory discussions concerning the facility resulted in an application to the state of Georgia for a Private Industry Solid Waste Processing permit. While the permit is still in the process of being finalized, the state has allowed the facility to operate under provisional entitlement. The second critical factor was the availability of an appropriate site at which to locate the facility. In the process of downsizing, CSXT was dismantling a number of buildings in its Old Savannah Yards, one of which was an old diesel shop. In its last few years of operation, it had been used for running repair. As such, the 300' X 100' building contained elevated track on which locomotives and other equipment could be brought into the building for repair. Repair pits allowed workers to service the units from below and cat walks permitted above ground repairs.

Viewed from a remediation perspective, the building was a very large cell with 8-inch reinforced concrete walls and floor and a clear height of 36 feet, all under roof. Furthermore, the building was located on a large clear site which contained several large concrete pads (previous building foundations).

The existence of a suitable site and building was crucial for the initiation and success of the project in that it provided a significant cost savings for the whole project. The decision was made to proceed with the project which would involve site and building renovation, development of a biological treatment protocol, and operation and management of the facility.

BUILDING DESIGN AND CONSTRUCTION

Preparation of the site involved only the removal of some debris and the installation of a fence and gate. The structure of the building was inspected and was determined to be sound. Major building modifications were limited to the demolition of an adjoining covered port and the replacement of some asbestos-containing siding. The majority of the work consisted of converting the operating portion of the building from a repair shop to a soil recycling facility.

Inside the building, all of the existing piping was removed and any piping or drains entering the floor or walls were sealed. Two cat walks running down the center of the building were removed, leaving the two side cat walks running the length of the building. Concrete ramps for moving equipment and storm drains to keep storm water out of the operations area were installed. Finally, the entire building was pressure washed and the floor was sealed with epoxy.

The building contains three repair pits that run the length of the building. These pits were designed as sloping troughs to collect any fluids that were spilled during the repair process. It was an easy task to convert these troughs into a leachate collection system by installing PVC drain pipes. In addition, the trenches are also used to provide air to the soil cell. Air injection piping was laid in the trenches next to the drain pipes. The trenches were then filled with gravel to floor level providing a flat surface for operations in addition to providing aeration and drainage.

OPERATIONS

Facility operations consist of logistics, remediation, and administration.

Logistics

Soil enters the facility by truck and is staged on-site on impermeable concrete slabs. The site has over 23,000 square feet of concrete within the fenced area and can accommodate several thousand yards of soil at any one time. Staged soil is covered with plastic to retain moisture and as a shield from the weather. Soil is moved on-site using a small tractor fitted with a loader bucket or backhoe/loader. Loading and moving 600-800 tons of soil for each treatment cycle is a major task for the operation of the facility. Prior to loading the treatment facility, the soil is mixed with a bulking agent and amended with nutrients. These tasks are accomplished at the same time using a hammer-mill type mixer. Inside the building, the mixed soil is spread evenly over the floor using the tractor.

Soil is removed from the building using the tractor and backhoe/loader and placed directly into trucks for transportation off-site or is again staged on site on a concrete pad. Due to logistical constraints, treated soil is removed from the site as soon as possible and preferably in the same operation as unloading the building.

Remediation

Remedial operations consist of maintaining optimal conditions within the facility for microbial activity. The air supply to the soil is maintained by blowing air through pipes in the trenches at a constant rate. The air is forced up through the gravel and permeates through the soil at a flow rate sufficient to ventilate the soil but not volatilize the petroleum contaminants. The blower is controlled by a timer which allows for intermittent operation if so desired.

The soil moisture is controlled by a sprinkler system mounted on the cat walks. The system is divided into eight individually controllable sections allowing the possibility of different moisture regimes for a single batch or multiple batches of soil under treatment at any one time. The leachate collection system provides a means of removing excess water from the system before irreparable damage is done to the soil under treatment. As leachate is collected in the troughs, it is pumped into a holding tank. It is expected that appreciable leachate will only collect if there is a failure in the watering system or if extreme weather conditions excessively water the soil. Any leachate that collects will be stored and sprayed onto the soil as moisture is again required.

Much of the operating system is controlled automatically by remote sensors and controllers connected to a computer system. The computer also automatically collects operating data which is stored for record and future analysis and reference.

Those parameters that are recorded are soil moisture, soil temperature, air temperature (inside and outside), relative humidity, wind speed and direction, and barometric pressure. All of the present readings can be accessed in real time from a remote location. The sprinkler system is controlled by the soil moisture measurements. Soil probes throughout the building record soil conductivity which is converted into soil moisture. The normal timed sprinkler cycle is interrupted if the soil moisture rises above a set point. Once the soil dries below this point, the sprinkling cycle resumes.

The soil is turned and tilled on a regular basis. This is accomplished with the tractor fitted with a bottom plow and a rototiller attachment.

Administration

Meticulous records of the entire operation are kept concerning the origin and fate of the soils treated and also the operations data as described above. Careful tracking of the origin, treatment, and fate of the soils are required for local, state and federal regulations. Each shipment in and out of the facility is carefully documented and recorded. Analytical results from each batch before, during and after treatment are noted and filed. Shipments requiring special handling or documentation are so noted. All of these records are sorted on the facility's computer system and can be accessed from any remote location via telephone link.

The operations data on temperature, moisture, etc. are automatically logged on the computer and stored for future use. It is anticipated that it will eventually be possible to predict, with accuracy, the progress and final result of treatment for any given batch of soil.

BIOLOGICAL CONSIDERATIONS

The soil recycling facility presented several specific challenges for the microbiology associated with the remedial process:

1. Production of a robust, inexpensive, and easily-produced inoculum.

2. Sustained, rapid activity.

3. Low remedial endpoints.

4. Ability to degrade mainly diesel fuel but have the ability to also handle other petroleum contaminants, appearing on a sporadic basis.

These requirements had to be met within the physical operational framework of the facility. The operating parameters that could be controlled were inoculum, soil amendments, nutrients, soil depth, tilling, moisture, and to an extent, aeration. The facility also has uncontrollable variables including temperature, wind speed and direction, relative humidity, and soil type. The challenge then, was to accomplish the overall goals of the facility with somewhat limited control of the operating parameters. A brief discussion of these parameters and variables follows.

Inoculum

Careful consideration was given to whether or not to add microbial cultures to the soil under treatment or to just stimulate the natural populations of microbes already present in the soils. Preliminary indications were that most of the soils that would be treated at the facility had large, diverse, and active microbial populations already in place and that remediation would proceed without the addition of any exogenous populations. These facts not withstanding, it was decided to augment the natural populations with a culture mix for three reasons: 1) for added insurance that the most appropriate microbial population would be present, predominant, and active; 2) to assure as rapid and complete remediation as was biologically possible; and 3) as an academic experiment to gain additional experience and information concerning soil remediation in a fixed facility.

Some of the "soils" that are present on CSX and other railroads' sites can hardly be defined as soil in the strict sense of the word. Fill materials of all sorts, ballast, debris, and ash are all common on railroad sites. Many of the soils contain a large percentage of foreign material. A concern was that this unnatural soil would be devoid of a microbial community or contain so few microbes that remediation would proceed slowly, if at all. In addition, even with natural soils, occasionally a high clay or sand content is encountered and the microbial populations are considerably reduced compared to a fertile field soil. In addition, in areas of extreme contamination, the toxic effects of the contaminants may have reduced or eliminated the natural populations.

A second concern, remediation to stringent endpoints, continues to be a challenge. Ideally, we would like to be able to reduce the petroleum contamination to below ten ppm. This will allow reuse of the treated soil as clean fill. However, to achieve this level in an economically feasible timeframe, it will be necessary to have every aspect of the facility optimized, including the microbial community.

To produce a robust and active inoculum, microbial cultures were chosen from ten typical CSX sites. These included active and inactive railyards, repair and refueling facilities, and several sites that had been leased by CSX to various fuel distributors and jobbers. The only common denominator among the sites chosen was a long history of moderate to extensive petroleum contamination. Soil samples were collected from each of the ten chosen sites and sent to our laboratory for microbial analysis. Several microbial isolates were chosen from each sample and evaluated for their petroleum-degrading potential and growth and activity characteristics. These characteristics were evaluated for their desirability in the soil recycling facility and seven cultures were chosen for inclusion in the facility inoculum. In addition, three isolates from our permanent culture collection known for diesel related characteristics were added to the inoculum mix to complement the CSX isolates. While the exact mixture remains proprietary, strains were chosen to be included based on the type of petroleum preferred as a substrate, the rate and extent of degradation, their competitiveness and survivability in a mixed community under the conditions existing at the facility, and their vigor and ease in handling. The inoculum was prepared by growing pure cultures of the ten microbes to a culture density of approximately 10^{11} cells per liter of culture. The cultures were then concentrated into a cell paste, combined and shipped live to the facility by overnight courier. On site, the cell paste was resuspended in a buffered solution and sprayed evenly onto the prepared soil in the facility. If possible, subsequent soil inoculation will be done by adding previously treated soil to the incoming soil thus perpetuating the original inoculum and eliminating the need for inoculation from cultures each time the facility is charged with soil. Over time, the petroleum-degrading microbial community will be selected and enriched in the soil by the specific conditions present at the facility. Due to the fact that most of the strains of microbes came from CSX soils, they are expected to compete well with the natural community. Since the individuals and their abundances will adjust and equilibrate, we will periodically reexamine the microbiology of the soils during and after treatment and make any adjustments needed. The data obtained from this study will be used to fine tune the inoculum if the incoming soil changes significantly, if a catastrophic event harms the soil under treatment, or if the facility is expanded or another facility opens.

Soil Amendments

Most of the soils encountered at CSX sites have a high clay content making the soils heavy and impermeable to water and air. If a heavy soil cannot be mixed with a more permeable soil coming in for treatment at the same time, a bulking agent is added to the soil prior to treatment. The agents that can be used include sand, vermiculite, peat moss, wood chips or trimmings, leaves or plant clippings, etc. In addition, several formulations of man-made materials are available. Generally, the bulking agent is added to provide increased moisture holding capacity and increased aeration. Cost and availability are more important in the decision for choosing a bulking agent than its effectiveness. All of the above agents are approximately equally effective from a microbial standpoint. Many of the natural agents may be unavailable locally and most of the artificial products are expensive. The final decision, therefore, may be a fait accompli. At the soil recycling facility, we have been using a mulch mix that is readily available locally and rather inexpensive in bulk. The mix contains a high percentage of untreated wood and wood fibers making it excellent for water retention and aeration. While the type of bulking agent may be predetermined for the specific site, the amount needed to be mixed with the soil is variable. Heavier impermeable soils obviously require more than light permeable soils but the exact mix will depend on facility conditions as well. For example, a forced aeration system may require less bulking agents than one with a passive aeration system. We have been experimenting with the soil to bulking agent ratio in order to optimize our results. Currently we are using approximately 20% bulking agent which has produced acceptable results so far.

Nutrients

The nutrients that are generally considered for addition to a bioremediation project are nitrogen and phosphate. Generally, site soils are poor in nutrients and additions must be made at the outset. Nitrogen and phosphate are routinely added in the Savannah facility. While the specific chemical formulation for the addition is proprietary, the chemicals are common agricultural fertilizers. One addition is made at the beginning of remediation, followed by monitoring of the remediation progress through soil analyses for the nutrients. The effect of periodic applications of nutrients throughout the treatment process is currently being evaluated.

Although not strictly nutrients, pH buffering agents are often discussed in the context of soil nutrients. These pH adjusting and buffering agents are required only when the pH of the incoming soil or the treatment process is less than about 5.5 or greater than about 7.5. Since the soils received so far have been slightly acidic, the initial pH has been adjusted with a one-time application of buffering agent at the time the soil is put into the facility.

Soil Depth, Tilling, Moisture, And Aeration
These factors are closely related and can be discussed as a group. Proper moisture levels and an adequate supply of air are essential for maximal microbial activity. Moisture is relatively easy to control and the facility's sprinkler system has already been described in detail. The biggest concern with moisture addition is adding too much water and having the system water log and become anoxic. Short of anoxic conditions there is a rather large range of moisture levels that will permit adequate, if not optimum, microbial activity. Aeration is a much more difficult parameter to control and to supply air evenly to a large mass of soil can be challenging. Forcing air through soil is superior to a passive air diffusion system even if the passive system contains very porous soils. To augment the air diffusion system the soil is kept bulked appropriately and limited to a the depth of 18 inches. If adequate tilling (completely turning the entire soil column) is available, deeper soil lifts can be tolerated. However, without specialized (and more expensive) equipment, tilling is normally restricted to 12-18 inches. In addition to aiding in aeration, tilling also mixes the soil which is beneficial to soil texture as well as microbial activity. This system responds well to tilling twice a week but the exact frequency for optimum results is still under investigation.

System Variables
Temperature is by far the most important controlling factor for environmental microbial activity. Generally, microbial activity doubles for every 10 Centigrade degree (18 Fahrenheit degrees) increase. Unfortunately, heating an outdoor soil pile is uneconomical unless some source of waste heat is readily available. Such is not the case at the recycling facility. The location of the facility, however, is in the Southeast where warm temperatures extend for most of the year and low (below freezing) temperatures are rare. Passive solar heating can be employed and steps can be taken to reduce heat loss from the soil. These ideas are currently under consideration but have yet to be implemented at the facility.

As with temperature, wind and humidity are also mainly controlled by the local weather conditions. Wind speed and direction and humidity primarily influence soil drying and, theoretically, soil ventilation. Unfortunately, little soil venting occurs from wind blowing over only the surface of the soil. If soil piles or windrows can be constructed economically, there can be some advantage to channeling wind through or across the site. At the recycling facility, since the building was already in existence, no consideration was given to positioning the building to take advantage of the prevailing wind direction. So far only generally flat soil surface has been utilized. In the future we hope to be able to construct windrows or channels to increase the amount of soil that can be treated at one time.

Soil types vary greatly with location. It is anticipated that most of the soil that is treated at the facility will be a silty clay, a rather dense and heavy soil. As already discussed, soil bulking through amendment is required at the facility. Even so, remediation proceeds much slower in these soils than it would in a less dense sandy loam. Increased remediation through soil blending can be successful if various types of soils are available and staged on site prior to treatment.

RESULTS

The soil recycling facility is still in the start-up phase but the preliminary results are impressive. Soils contaminated with gasoline only and diesel fuel only have been successfully treated. Within a four (4) week period, soils starting with a total petroleum hydrocarbon content of 2,000 to 3,000 ppm were treated to less than detection (gasoline) and 100 ppm (diesel fuel). As experience is gained in operating the system, it is anticipated that the treatment time will decrease and that the endpoint for diesel fuel contaminated soils will approach the goal of 10 ppm.

Table 8-1. Facility Costs.

Fixed Costs

Building/facility rehabilitation	$329,701
Engineering/Construction	
Initial Microbe Isolation	$20,000
Process Documentation	$14,000
Total Fixed Costs	**$363,701**

Cost of Debt

Project Life Cycle	10 Years
Project Cost of Money	6%
Annual Debt Service	**$49,415**

Variable Costs

Annual Tons of Soil	9,648
Monthly Tons of Soil	804
Total Fixed Cost/Ton	$5.15
Inoculum/Ton	$2.00
Analytical/Ton	$3.60
Operations and Maintenance/Ton	$7.22
Total Unit Cost	**$17.97**

FINANCIAL CONSIDERATIONS

At the projected optimum monthly capacity, the soil recycling facility will treat petroleum-contaminated soil at less than $17.97 per ton. For comparison, the local market conditions for other acceptable forms of remediation have established a lowest base cost of approximately $28.00 per ton. The net savings to CSXT, therefore, when the facility is fully and optimally operational is greater than $10 per ton. Recognizing that the facility would not necessarily run at optimum capacity, a spread sheet was constructed showing a range of 600 to 1,500 tons of soil a month and how the various costs would fluctuate. Table 8-1 shows the breakdown of cost associated with operation of the facility. The fixed costs include the debt service required for the building and startup costs allocated assuming a capacity of 804 tons of soil per month and a 10-year facility life.

The operations and maintenance portion of the costs represents the largest component of the total cost. We have also included the cost of the chemical analyses ($3.60 per ton) although strictly speaking, this is not a remediation cost. Due to the high fixed costs involved, the cost per ton is facility capacity sensitive, as shown in Figure 8-1. With a 50% increase in throughput, the price per ton can be reduced to below $16. Figures 8-2 thru 8-5 illustrate the cost breakdown of treatment as the facility capacity increases. As expected, the operations and maintenance component increases disproportionately while the fixed facility portion decreases.

CONCLUSIONS

This paper presents a starting point in the total evaluation of a fixed, ex-situ bioremediation facility for treating petroleum-contaminated soils. Even though still preliminary, the results conclusively show that soil can be treated in this facility in a cost effective manner. The initial results indicate a minimum saving of 35% over the least expensive alternative. It is anticipated that additional savings can be realized by CSXT once the system is fully analyzed and fine tuned. Future questions to be answered include:

- What is the optimum output of the facility?

- What operations processes can be streamlined to allow for cost savings?

- What are the most important variables in the process?

- Can the microbiology of the system be manipulated to reduce the treatment time or ultimate remedial endpoint?

The development of this type of effective and economical soil remediation technology will help keep the railroad industry and CSXT "Environmentally On Track."

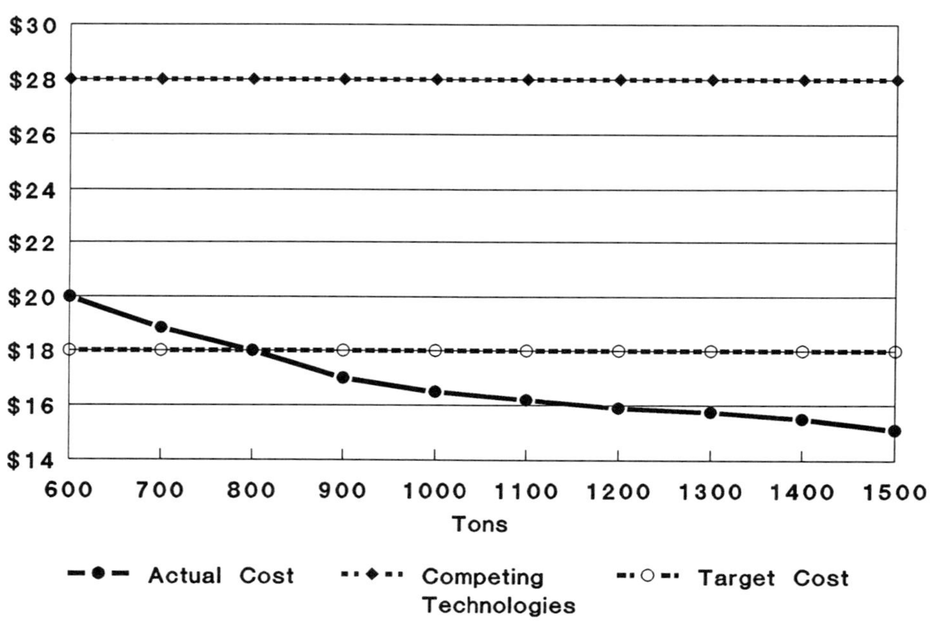

Figure 8-1. Comparison of unit costs for ex-situ and alternate technologies.

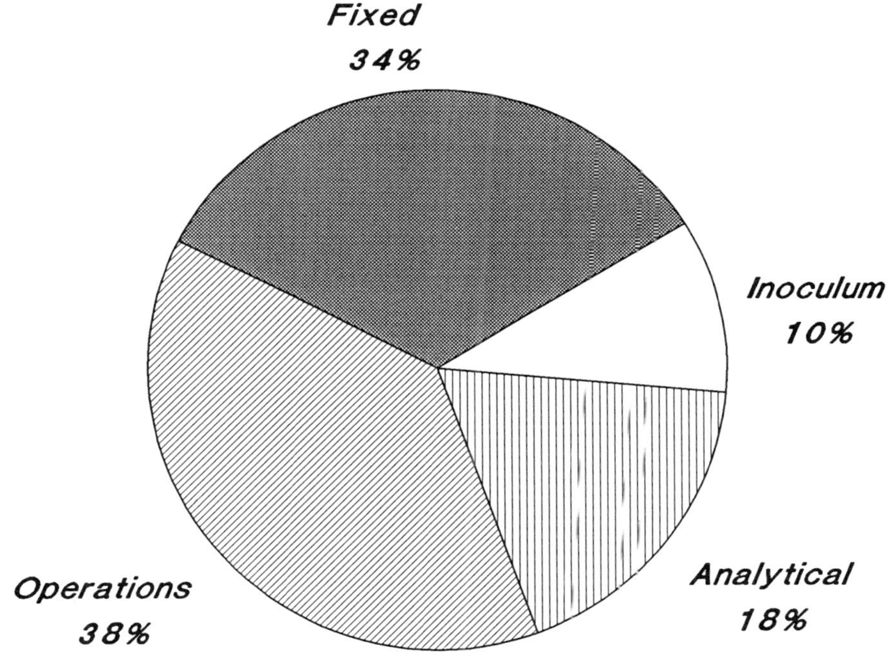

Figure 8-2.   Cost breakdown by major category at a volume of 600 tons per month.

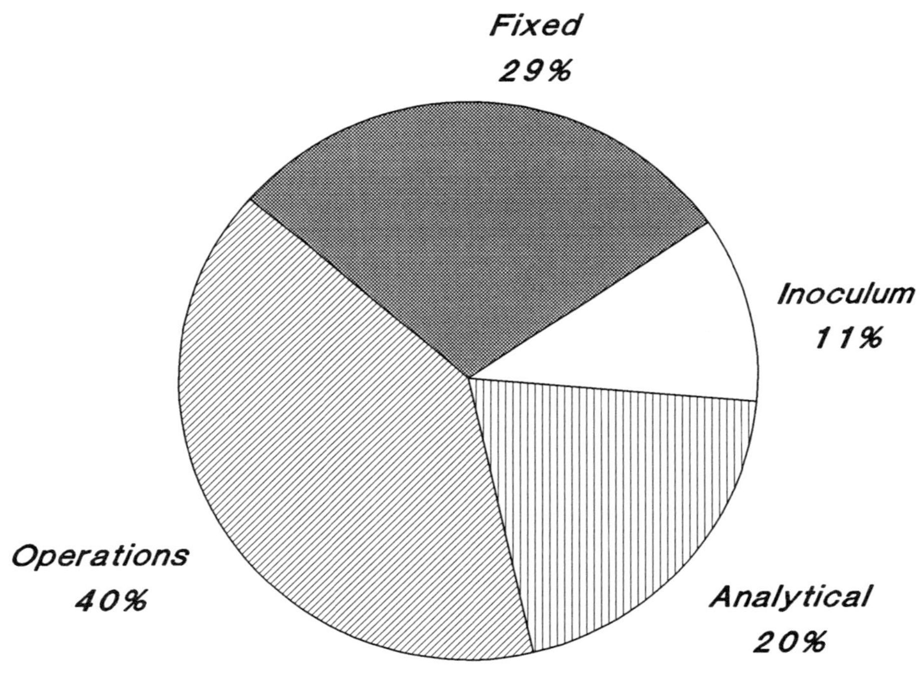

Figure 8-3.	Cost breakdown by major category at a volume of 800 tons per month.

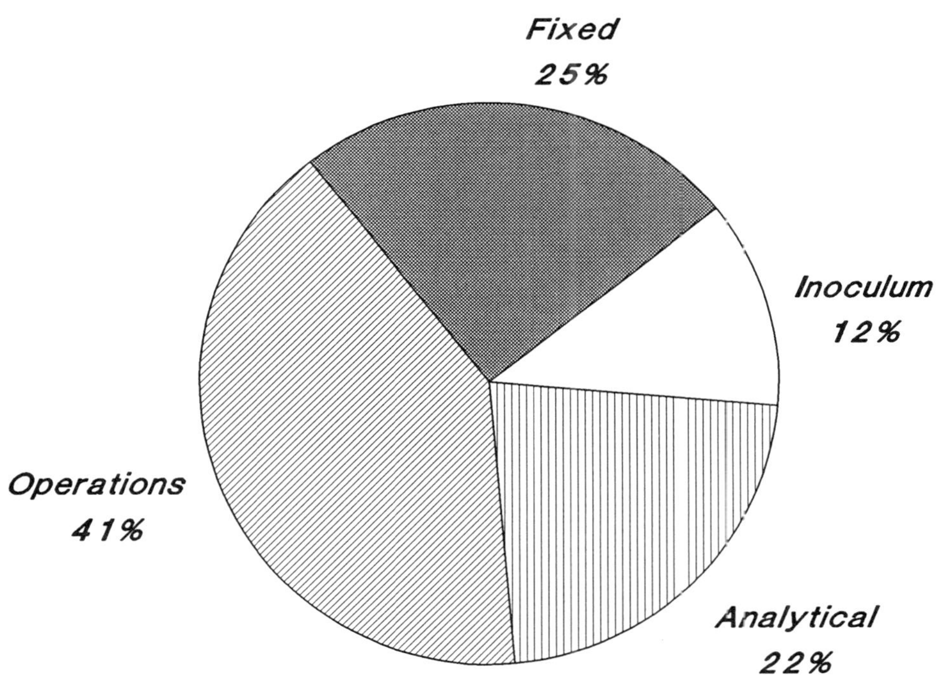

Figure 8-4. Cost breakdown by major category at a volume of 1.000 tons per month.

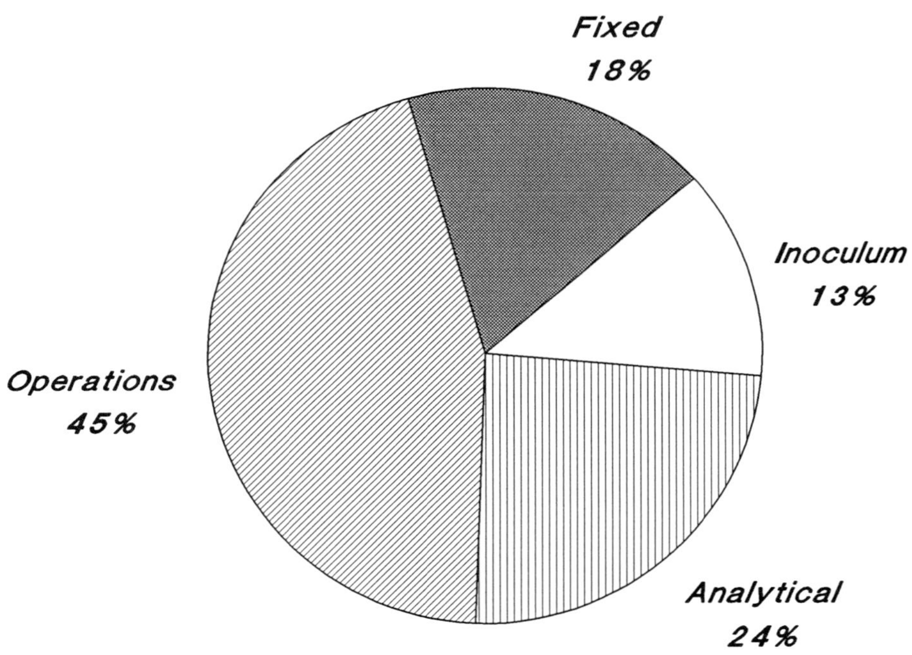

Figure 8-5.   Cost breakdown by major category at a volume of 1,500 tons per month.

Questions and Answers:

Q. A quick question about the type of contaminants; on occasion do you get soils contaminated with PCBs?

A. No, we're prescreening prior to putting anything into the facility. We have pretty stringent criteria set up for what we will accept and won't accept, PCBs being one that we will not accept. Also we will not accept soils with certain metals.

Q. Second question. The analysis work that is done while a batch is in there; how do you do your analysis? Is it sent off-site and what do you actually test for?

A. We are running 8015s, diesel range organics (DRO) and gasoline range organics (GRO). It is sent off-site to Savannah Labs which is down the street from us. The analysis gives us an indication of our TPHD. The ultimate criteria for accepting soil are the TPH and the BTEX values. The State of Georgia has agreed that if we can get it under a hundred parts per million TPH it can be used as daily cover for landfill. We have an agreement with BFI regarding this. If we can get it under ten ppm we can put it back into the site that it came from as clean soil. We are approaching a level of 10 ppm right now.

Q. What temperature do your soils get to in the treatment facility?

A. Right now the temperatures are remaining at a fairly constant level, between 72 and 76 degrees fahrenheit.

Q. Do you ever get up to the 80 - 90 degree range?

A. Air temperature, yes, but not the soil temperature. It's been pretty much constant over the whole life of the facility so far.

Q. Do you have any information or do you think that the microbial population will work faster if your soils are that warm?

A. Yes, we believe that if we can increase the temperature by ten degrees centigrade, we can double the population.

Q. Does anybody have any information that shows that effect in the 90 to 100 degree range? I'm operating a facility in the desert and my concern is the high temperatures and also drying.

A. What moisture content are you at?

Q. Well I'm trying to stay in the range of 50% of the holding capacity of the soil but that's hard to do in the desert. I just wondered about that temperature and also the drying impact. Do you know at what point soils get dry enough to inhibit some of that activity?

A. We have a system in place which you didn't see and I can show you later if you're interested. We have soil temperature probes throughout the whole building that go into the controller and when the soil moisture content is reduced below 40% the computer system automatically turns on the sprinkler and the sprinkler automatically comes on inside the building.

Q. Do you find that at times you run into spikes in your analysis where your TPHs seem to be growing, where you've degraded down to a certain level and all of a sudden you get a spike and it comes back up?

A. We could have a very long discussion about that because of EPA Method 418.1 versus 8015, and I could take an hour talking about that, but yes, we have had some instances of that, but for the most part by now using the 8015 we've reduced some of our spikes.

Q. I am Sara Tremaine, of Hydrosystems, I missed in your budget whether you included transportation charges for the soils?

A. No, transportation charges were not included.

Q. How do you think that would figure into your 18 dollars a ton cost figure?

A. To bring contaminated soils from within the country to the facility costs up to $3.65 per ton in transportation charges. The nearest commercial facility that would accept these types of soils is the Kadish incinerator in Brunswick, Georgia Transportation costs to that facility would be close to $7.00 per ton.

Q. You're providing enough air underneath your soil to promote biological activity, but not enough to use volatilization as a major mechanism for product removal. How do you go about balancing that?

A. About balancing the air flow?

Q. Yes, so that you're not volatilizing product rather than biodegrading it?

A. We are blowing air at such a low rate, about 10 CFM that it really doesn't matter; we're just adding enough. I believe that there's enough air getting in there just by the amount of turning that we're doing that we could actually turn that system off. I was persuaded by our consultants that we needed to put these in, but I don't believe that it's actually aiding right now, at all.

Q. Do you monitor the air in the work space over the soils?

A. No, we do not.

Q. I'm just curious what level you try to maintain your nutrients at?

A. Good question. In the last run through we put in 400 tons of soil and we added 800 pounds of nitrogen and we'll expect to see about a 30 to 40% decrease in nitrogen when the soils reach their treatment end point.

CHAPTER 9

From Laboratory Studies to Full-Scale Bioremediation of Diesel Contaminated Soils with High Levels of Chlorides at a Superfund Site

David Clark, P.E.
The Atchison, Topeka and Santa Fe Railway — Topeka, Kansas

Timothy P. Wippold, P.E.
Radian Corporation — Houston, Texas

INTRODUCTION

Radian Corporation (Radian) is conducting the phased remediation of a hydrocarbon contaminated, National Priority Listed (NPL) Superfund site located in Clovis, New Mexico. Figure 9-1 identifies the playa lake system with respect to the railroad facility. The Atchison, Topeka and Santa Fe Railway Company (Santa Fe) has operated a railyard just north of Santa Fe Lake since the early 1900's. Petroleum hydrocarbons were introduced into the playa lake system primarily from wastewater discharges from hopper car washing operations; also contributing contaminants to the Santa Fe Lake site were storm water runoff and wastewater discharges from boiler blow downs, sanitary sewers, and oil/water separators. After evaluation of several remedial options, biodegradation of contaminated soils and lake bottom sediments was chosen as the best practical alternative and placed into the Record of Decision (ROD) in September 1988. In order to fully optimize conditions for full-scale bioremediation at the site, Radian designed and implemented various laboratory and field studies. An initial treatability study was conducted to determine if the hydrocarbon contaminants in both soils and lake bottom sediments could be biodegraded using indigenous microorganisms. Since chlorides were found to inhibit biodegradation with indigenous microbes, a subsequent laboratory study determined if commercial chloride-tolerant microbes could be used to biodegrade the contaminants in the lake bottom sediments. A field study, used as a forerunner to full-scale biodegradation, was conducted to determine how hydrocarbon degradation in soils would proceed in field conditions. This study would pave the way for optimizing factors influencing biodegradation and allow the project team to investigate alternative methods for bioremediating the lake-bottom sediments. Full scale bioremediation would then, systematically incorporate the results of the former treatability studies into a complete and viable treatment plan.

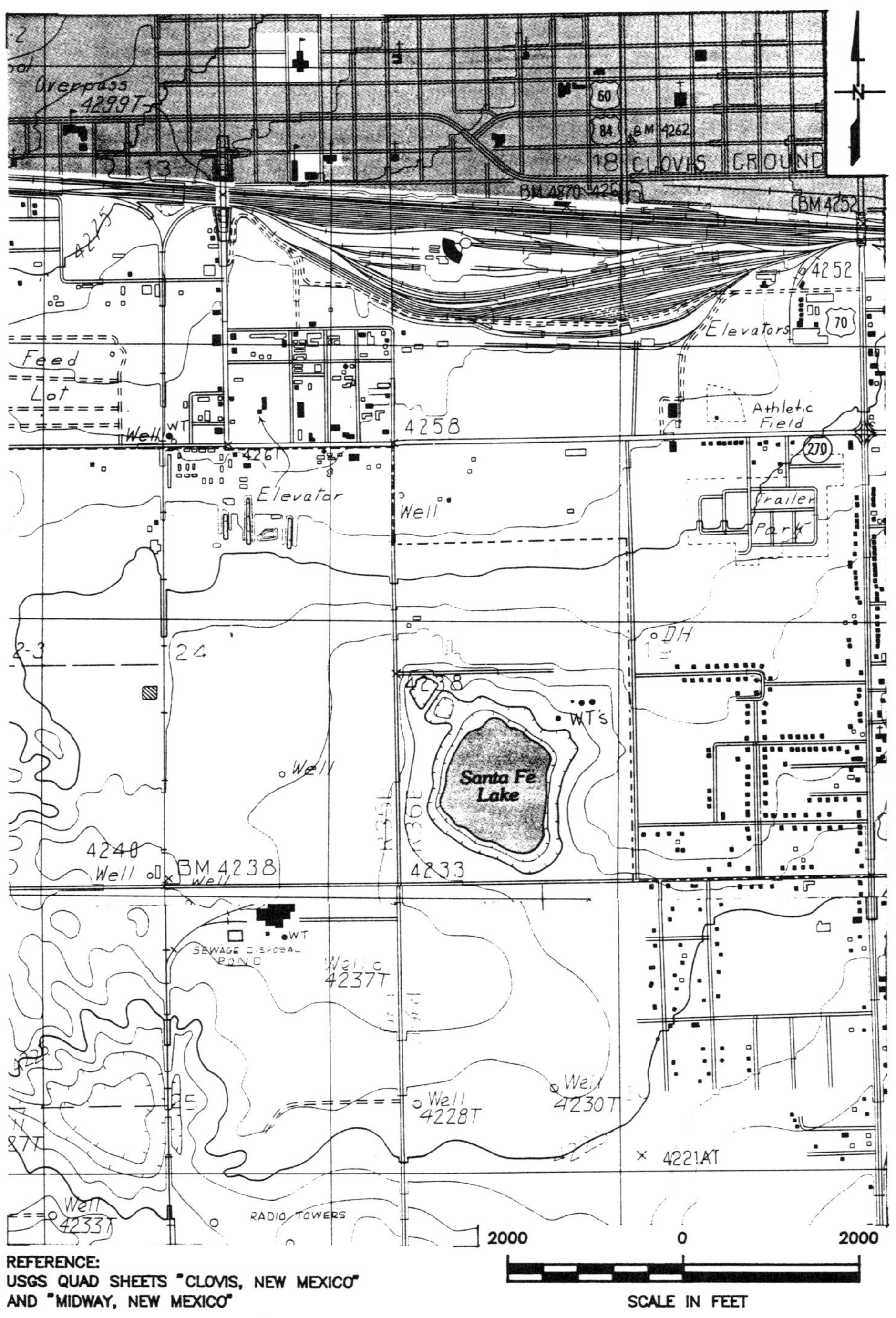

Figure 9-1.   Santa Fe Railyard and Sante Fe Lake Site Map, Clovis, New Mexico.

SITE BACKGROUND

Located one mile south of the Santa Fe railyard in Clovis, New Mexico, Santa Fe Lake, a natural playa lake, receives runoff from surrounding areas of approximately 810 acres. It has been estimated that at one time this playa lake covered an area of 31 acres and was approximately 9 feet in depth. Figure 9-2 illustrates the estimated historical high water mark, elevation 4,217, for Santa Fe Lake. Evaporation is the primary method of water loss for the Santa Fe Lake since the playa lake has no natural outlet. By 1988, the lake area had decreased to 16 acres. This volume reduction caused by evaporation has further impacted the lake site by causing a high dissolved solids concentration in lake bottom sediments and water.

Soils in the Santa Fe Lake area consist primarily of fine sandy loams and fine loamy sands except in the bottom of playa lakes where a clay layer exists. The Ogallala Formation runs approximately 250 to 300 feet underneath the playa lake surface. The Ogallala Formation is a thick sequence of alluvial deposits composed of sand, silt, clay, gravel, and caliche accumulations. The Ogallala aquifer is a primary source of water in this region.

Prior to 1962, Santa Fe Lake received wastewater discharges from boiler blowdowns, sanitary sewage, and oil/water separators at the fueling racks. During this time period, the estimated wastewater discharge ranged from 40,000 to 60,000 gallons per day. In 1962, a hopper car washing facility was constructed. The addition of the hopper car washing facility caused a marked increase in wastewater discharges to the lake site. It is estimated that from 1962 to 1975 the wastewater discharge averaged 100,000 gallons per day. From 1975 to 1979, hopper car washing operations peaked at 145,000 gallons per day. In 1982, the hopper car washing facility was closed and other wastewater discharge activities were minimized; the present estimated discharge is 8,000 gallons per day.

The hopper car washing facility was the principal contributor of wastewater at the Clovis facility. Hopper cars which had formerly contained potash, borate, cement, fertilizer, grain, coke, and other bulk commodities were cleaned at the railyard facility. The hopper car washing facility was constructed so that heavy solids removed from the hopper cars would remain on the wash platform; these solids were later removed by an endloader and a dump truck. The lighter, settleable solids and floating materials were retained in adjacent settling ponds. The final discharge from this facility, which contained high concentrations of dissolved salts, were piped to Santa Fe Lake.

Prior to 1983, the United States Environmental Protection Agency (EPA) sampled and analyzed the lake surface water, lake bottom sediments, and the groundwater from a nearby well. Analysis from this sampling event revealed the presence of cyanide, chromium, cadmium, and lead. The EPA further speculated that the permeability of the soil underlying Santa Fe Lake may be such that these constituents could percolate to the groundwater. The Santa Fe Lake site was placed on the National Priority List (NPL) based upon the presence of cyanide and its potential for leaching to the Ogallala aquifer.

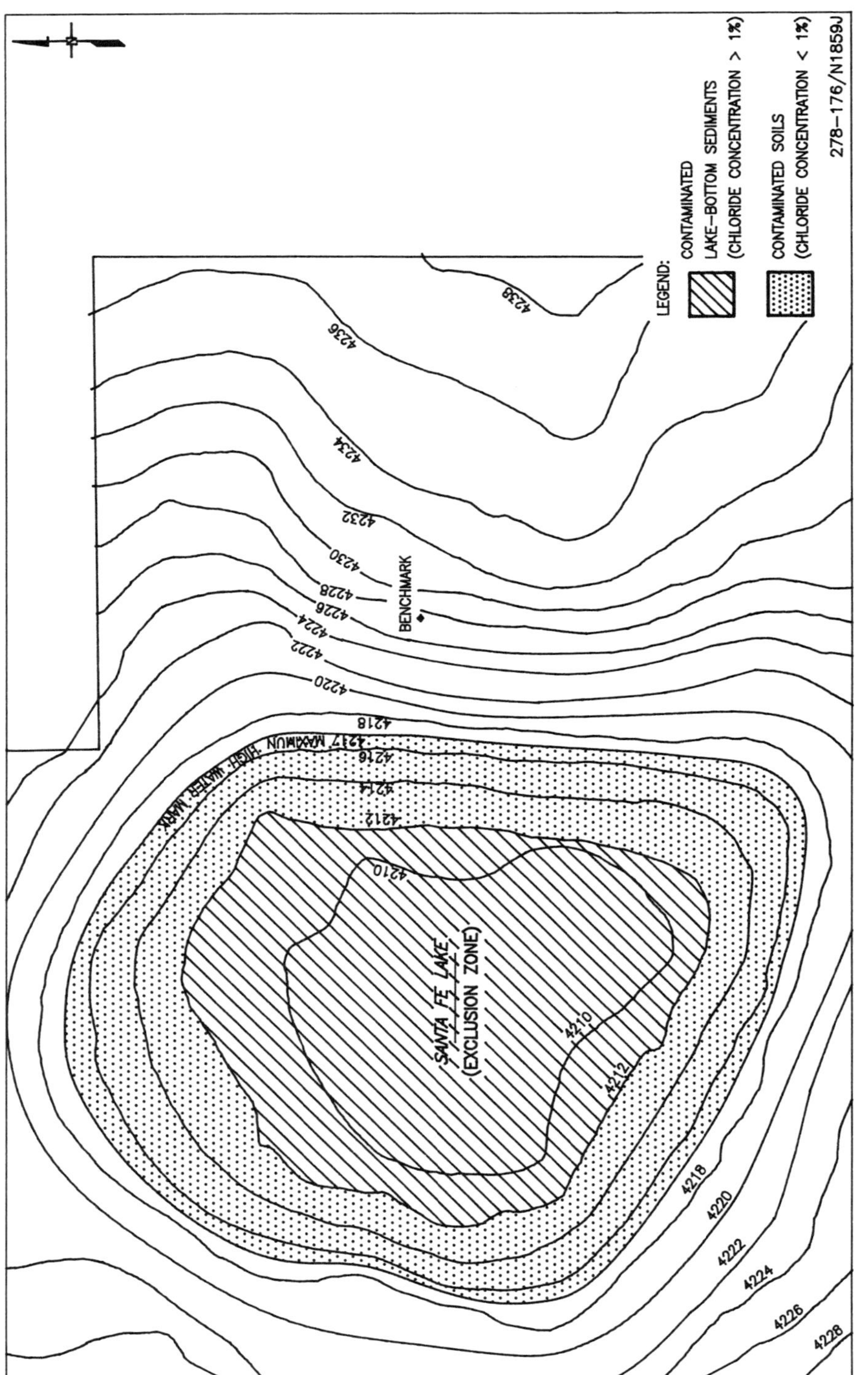

Figure 9-2.   Sante Fe Lake Elevation Contours Map.

For a period of two years, Santa Fe sampled several nearby wells and submitted their findings to the EPA. During this period, none of the constituents responsible for placing the Santa Fe Lake site on the NPL were detected; the previous findings were then assumed to be a laboratory error.

A Remedial Investigation/Feasibility Study (RI/FS) identified the nature and extent of contamination in the groundwater, lake bottom sediments, and soils surrounding the lake. The RI/FS developed cleanup criteria based upon health risk assessments and recommended remediation alternatives for the lake site. As a result of the extensive RI/FS, an integrated system addressed a plan of action that would achieve remediation while minimizing costs. The integrated system included the following:

- Construction of a perimeter fence;

- Construction of a run-on control dike around the contaminated area;

- Construction of a spray evaporation system within the diked area to evaporate the lake water;

- Construction of an on-site biodegradation cell;

- Excavation of the lake bottom sediments and placement in the biodegradation cell for treatment;

- *In situ* biodegradation of contaminated soils; and

- Capping of the biodegradation cell.

After the EPA approved the proposed remediation plan, a ROD was issued in September 1988. All surface water runoff and groundwater from the Clovis railyard facility fuel recovery system would be managed in a trench behind or outside of the run-on dike. In order to differentiate between the lake soils and the lake bottom sediments, the following definitions were imposed:

- "Contaminated soils" are defined as firm and coarse grained materials with a total petroleum hydrocarbon (TPH) concentration >1,000 ppm; while

- "Lake bottom sediments" are defined as soft and fine grained materials that generally possess elevated concentrations of chlorides.

These definitions were imposed in order to address each medium based upon its unique characteristics. Figure 9-2 locates the interface between the lake bottom sediments and the contaminated beach soils.

BIOMETER TEST PHASE/BIODEGRADATION STUDY

Selection of bioremediation as the preferred treatment method required careful planning in order to implement this treatment method on a large scale. The objectives of the initial treatability study were as follows:

1. To determine the overall landfarm treatability potential for the Santa Fe Lake site;

2. To estimate optimum lake bottom sediment loading; and

3. To determine the biodegradation rate of the contaminant of concern.

In-situ biodegradation is a biological treatment process that improves or stimulates the microbes metabolic capabilities. Certain microbes, under nontoxic conditions, are capable of ingesting hazardous organic residues that are deleterious to humans. Factors that may prevent the biodegradation process are as follows:

• Chemical concentrations toxic to microorganisms;

• Insufficient microbial populations;

• Inadequate type of microbes;

• Acid rich or alkaline rich environment;

• Lack of nutrients such as nitrogen, phosphorus, and other trace elements;

• Unfavorable moisture conditions (too wet or too dry); and

• Lack of oxygen or other terminal electron acceptors.

The biodegradation potential and rate of degradation of petroleum hydrocarbons in the contaminated soils and lake bottom sediments were evaluated based upon a series of studies. The initial bench scale treatability study provided pertinent information on lake site soils and lake bottom sediments biodegradation potentials and their rates of degradation. Biometer flask tests were used to estimate the biodegradation potential of the contaminants in the soil while a biodegradation study provided Radian with degradation rate information. The biometer flask test estimates the biodegradation potential of a contaminant in soil by measuring the soil bacteria respiration, and CO_2 evolution, from the flask system. By monitoring the test systems on regular intervals for contaminants of concern, rates of hydrocarbon degradation were determined.

Biometer Flask Study
A series of biometer flask tests were performed to analyze the relationship between biomass concentration and microbial activity specific to the Santa Fe Lake site. Soil types and lake bottom sediment loading were varied while the biomass spike was held constant for each test.

Return sludge provided by a Houston, Texas industrial wastewater treatment facility was used as the biomass spike material. This industrial wastewater treatment facility receives waste from area refineries and petrochemical plants; therefore, the microbial populations present are broad in spectrum and very active for degrading hydrocarbon contaminated wastes. The flask tests were set up in duplicate for the scenarios illustrated in Table 9-1. Care was taken to optimize environmental conditions for the microbes by adding the necessary nitrogen, phosphorus, and moisture. A gentle stream of compressed air was passed into the flask system through an ascarite filter to provide carbon dioxide free air to the system. Flasks were placed in a dark room at ambient temperature and removed at prespecified time intervals to measure the evolved carbon

Table 9-1

Clean Soils	5% Lake Sediments Clean Soils	10% Lake Sediments Clean Soils	20% Lake Sediments Clean Soils	Contaminated Soils
0.5% biomass	0.5% biomass	0.5% biomass	0.5% biomass	0.5% biomass
1.0% biomass	1.0% biomass	1.0% biomass	1.0% biomass	1.0% biomass
2.0% biomass	2.0% biomass	2.0% biomass	2.0% biomass	2.0% biomass

dioxide and to re-aerate the systems with carbon dioxide free air. A dark room was necessary in order to minimize the possibility for photodegradation by both biological and chemical processes. The flask test was monitored for a period of 19 weeks.

Results of the carbon dioxide evolution flask tests are as follows:

1. The contaminated soils had the highest levels of carbon dioxide evolved indicating a higher level of biological activity.

2. The control, uncontaminated soils, generally were higher in biological activity than the soils loaded with lake bottom sediments.

3. There appeared no apparent trends for increasing the concentration of lake bottom sediments to uncontaminated soil. In general, the lake bottom sediment loaded soils yielded lower carbon dioxide evolution rates than either the contaminated soil or the study control. It can be concluded that the lake bottom sediments appear to have an inhibitive affect on biodegradation.

Definite trends were observed for both the contaminated soils and for the entire study. Figure 9-3 graphically depicts differences among the biometer flask study specimens. Likewise, similar results were observed in Figures 9-4 and 9-5. Comparing Figures 9-3, 9-4, and 9-5 indicate that an increase in biomass concentration corresponds to an increase in the evolution rate.

In conclusion, this study revealed that the contaminated soils appeared to have a higher potential for biodegradation than the lake bottom sediments based upon the flask tests. Also, the environment created by the lake bottom sediments inhibits biodegradation indicating toxic conditions for the microbial population.

Biodegradation Study

The biodegradation study demonstrated how quickly the soils and lake bottom sediments would degrade. The biodegradation study was conducted as follows:

1. Biomass - 0.5%, lake sediments, uncontaminated soil - 10 percent;

2. Biomass - 0.5%, contaminated soil; and

3. Biomass - 0.5%, uncontaminated soil (control).

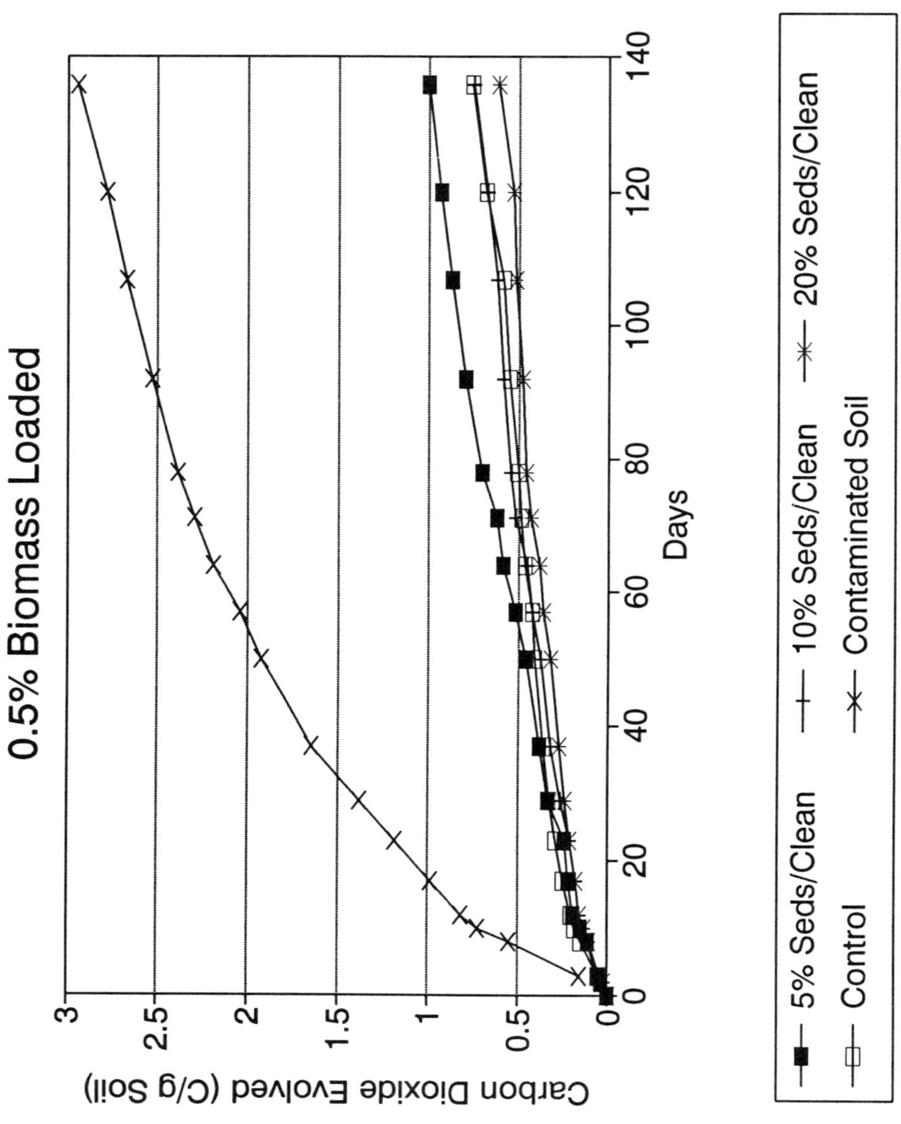

Figure 9-3. Carbon Dioxide Evolution Rates (0.5% Biomass Loaded).

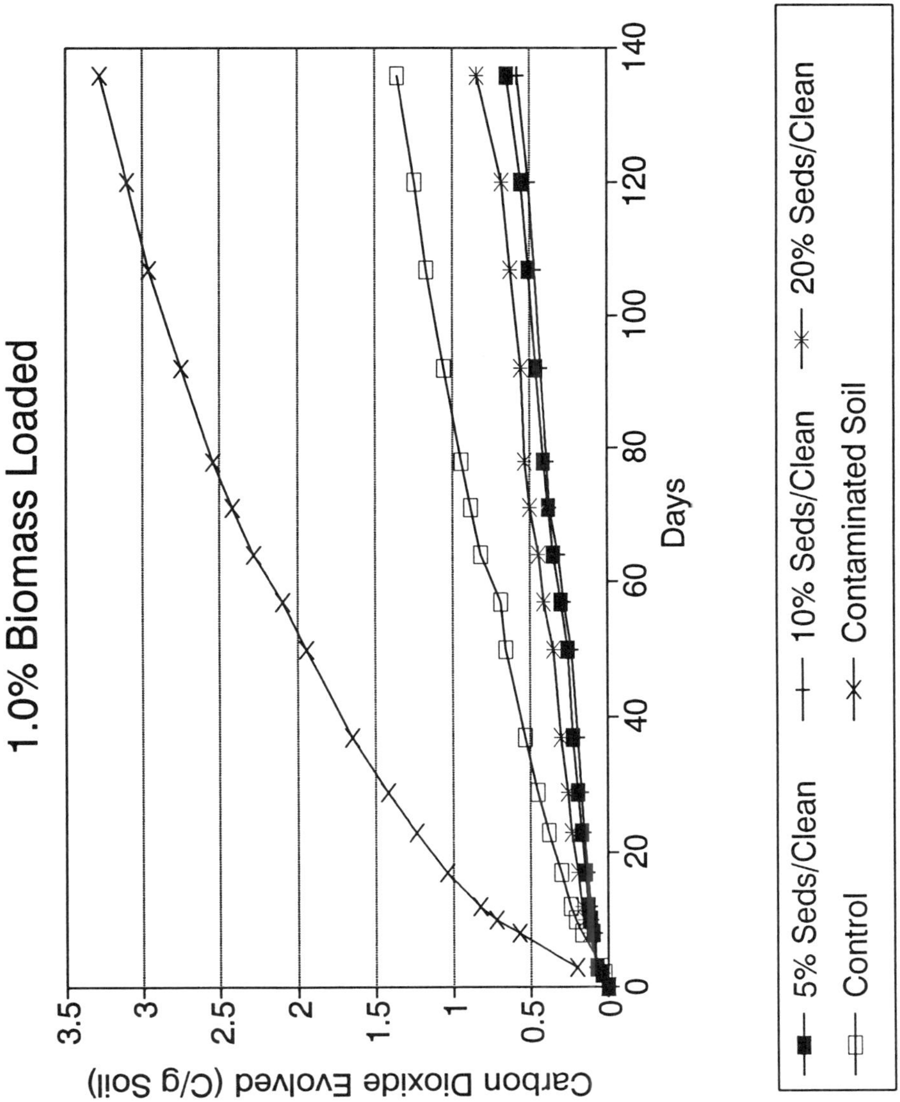

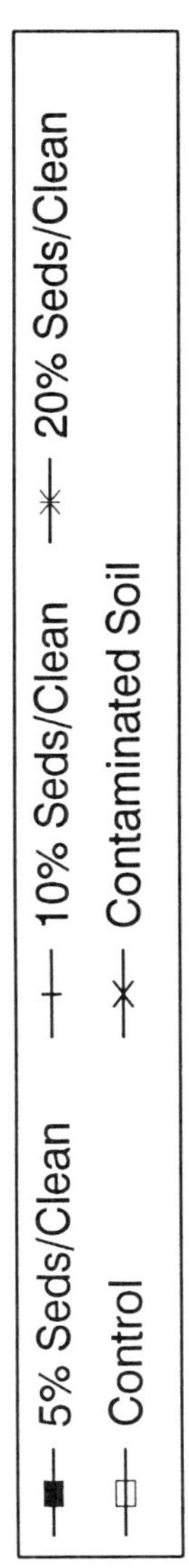

Figure 9-4. Carbon Dioxide Evolution Rates (1.0% Biomass Loaded).

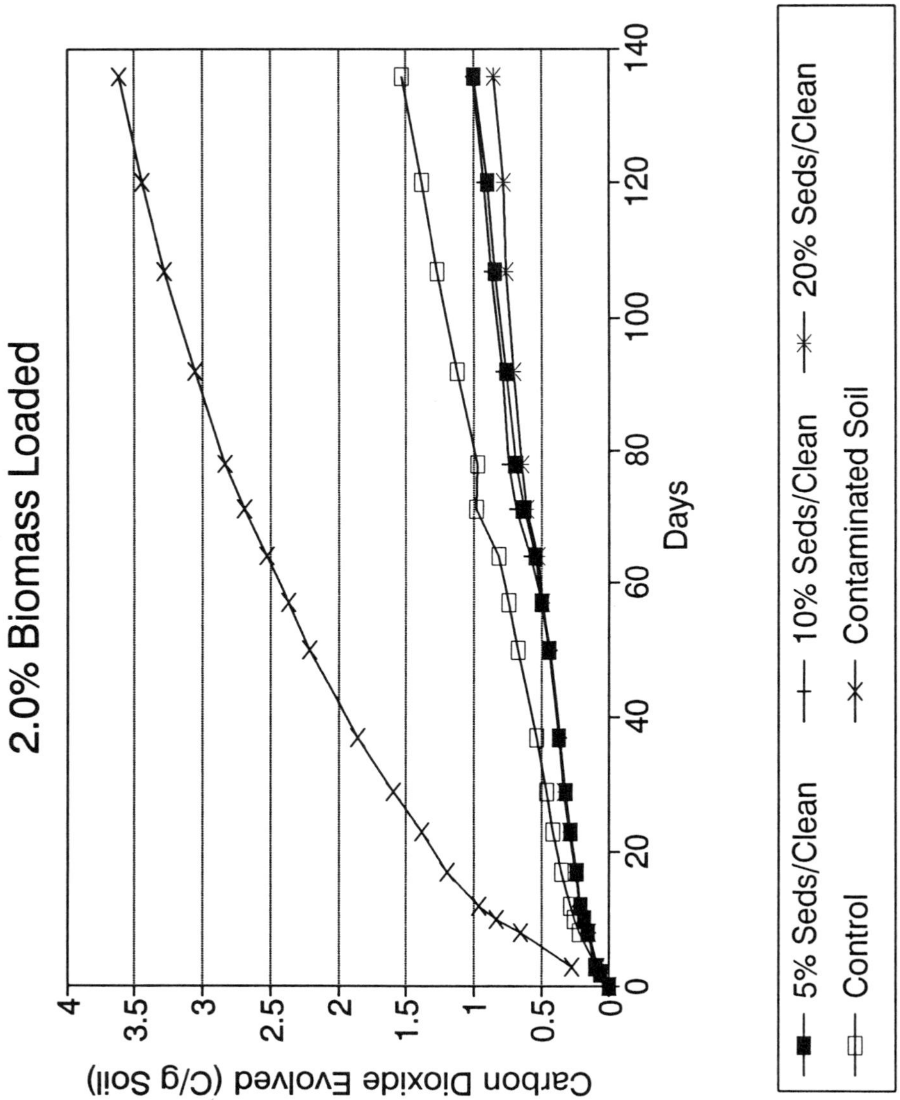

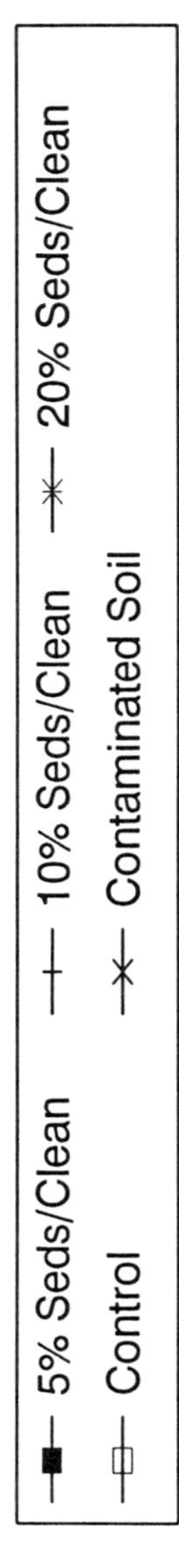

Figure 9-5. Carbon Dioxide Evolution Rates (2.0% Biomass Loaded).

Each test, with the exception of the control, was performed in triplicate. The control was performed in duplicate. At prespecified time intervals, the test chambers were sacrificed for analysis. Each system was analyzed for total organic carbon (TOC), oil and grease, volatile organic compounds, and semi-volatile organic compounds. The time period for the biodegradation study was 126 days. Test chambers were optimized by adding moisture, nitrogen, and phosphorus.

Several conclusions were drawn at the end of this study. Those conclusions are as follows:

1. The lake bottom sediments are not amenable to land treatment;

2. Fifty to 65% of oil and grease removal takes place after approximately four months using a biomass spike of 0.5 percent;

3. The contaminated soils are amenable to land treatment; and

4. The volatile and semi-volatile organics are not present at significant levels in the samples.

Once chloride levels were identified as the possible sediment biodegradation limiting parameter, another experiment was conducted to determine if chloride levels did, in fact, have an inhibitory affect on the biodegradation of lake bottom sediments. The biometer flask test and the biodegradation study in general confirmed that the beach soils had good biodegradation potential while the lake bottom sediments exhibited poor biodegradation potential, relatively speaking. A sampling episode determined that the soils had average chloride concentrations of approximately 0.8% while the lake bottom sediments had average concentrations of approximately 2.4% on a wet weight basis. According to the literature, chloride concentrations of 1% tend to create an adverse environment for biodegradation.

In the Santa Fe Lake site system, a 1% chloride content corresponds to the 4,212 contour line. Generally speaking, all soils outside of the 4,212 contour line are coarse-grained firm soils with a chloride content <1% while all soils inside the 4,212 contour line are very wet, soft and fine-grained lake bottom sediments with a chloride content >1 percent. This information created a definite boundary for *in situ* biodegradation and became a permanent reference for establishing physical works that would ultimately support full scale bioremediation. Figure 9-2 locates the 4,212 contour on the Santa Fe Lake elevation contour map.

FIELD PLOT STUDY

A two-phased field plot study was conducted to determine biodegradation performance potential of contaminants of concern under field conditions. The objective of Phase I was to measure the *in situ* biodegradation of hydrocarbons in the Santa Fe Lake site soils while acclimating soil microorganisms for use during Phase II. The objective of Phase II was to determine if the hydrocarbons in the lake-bottom sediments could be biodegraded using acclimated microorganisms developed in soils treated during the previous field plot study phase.

A plot is defined as a 15- by 15-foot area located just outside of the 4212 elevation contour line (contaminated beach soils). Four plots were established and identified as such: FP1, FP2, FP3, and FP4. Figure 9-6 locates and describes the field plot study area. Each plot received bioseed, nutrients, water, and cultivation for a period of eight weeks. Samples were taken throughout this testing phase. At the end of Phase I, an additional plot, FP5, was created as a control for Phase II. In Phase I, all plots were treated equally. However, in Phase II each plots role differed slightly. FP1, FP2, and FP3 received 10%, 25%, and 40% lake bottom sediments by weight, respectively. FP4 was not tampered with in order to determine the long term effects of *in situ* bioremediation under Phase I conditions. FP5 became the true control for the study since its microbial population would not be enhanced by the addition of bioseed. All other parameters including, cultivation, irrigation, and nutrient loading remained identical for all plots throughout the entire biodegradation study.

The field plot study findings for beach soils are as follows:

1. Approximately 70% to 80% degradation of TPH in soils was attained by *in situ* biodegradation within an the eight week period;

2. Beyond eight weeks, TPH biodegradation in the beach soils was found to be insignificant, relatively speaking; and

3. The C:N:P ratio of 50:2:1 (Phase I) versus 100:1:0.1 (Phase II) did not significantly effect the biodegradation of the beach soils.

Lake bottom sediments can be successfully biodegraded by mixing treated soils containing acclimated microorganisms with lake bottom sediments. The field plot study refuted earlier claims about the biodegradability of the lake bottom sediments at the Santa Fe Lake site. The initial treatability study findings reported that the lake bottom sediments were not amenable to land treatment. However, several important factors were not explored at the time of that test. For instance, the biometer flask test indicated that the contaminated soils at the Santa Fe Lake site were more biodegradable than uncontaminated soils under the same conditions. This suggests that the environment for microbial growth was better in the contaminated soil as opposed to the lake bottom sediments mixed with uncontaminated soil due primarily to the presence of substrate (food source). Also, the addition of lake bottom sediments to treated soils in the field plot study probably diluted the soil chloride concentration to "non-toxic" or "not-as-toxic" levels for the entire system. TPH degradation of sediments was achieved within a 10 week period. Each lake bottom sediment loaded plot experienced a significant decrease in TPH concentration in this study. However, workability becomes extremely laborious at 40% by weight lake bottom sediments.

BIODEGRADATION SITE PREPARATION

In order to meet the objectives of the Record of Decision (ROD) issued by the EPA, a multi-phased approach was planned and implemented. Phase I was the construction of run-on dike that would encircle the contaminated lake area. Phase II was the construction of an irrigation and spray evaporation storm water management system.

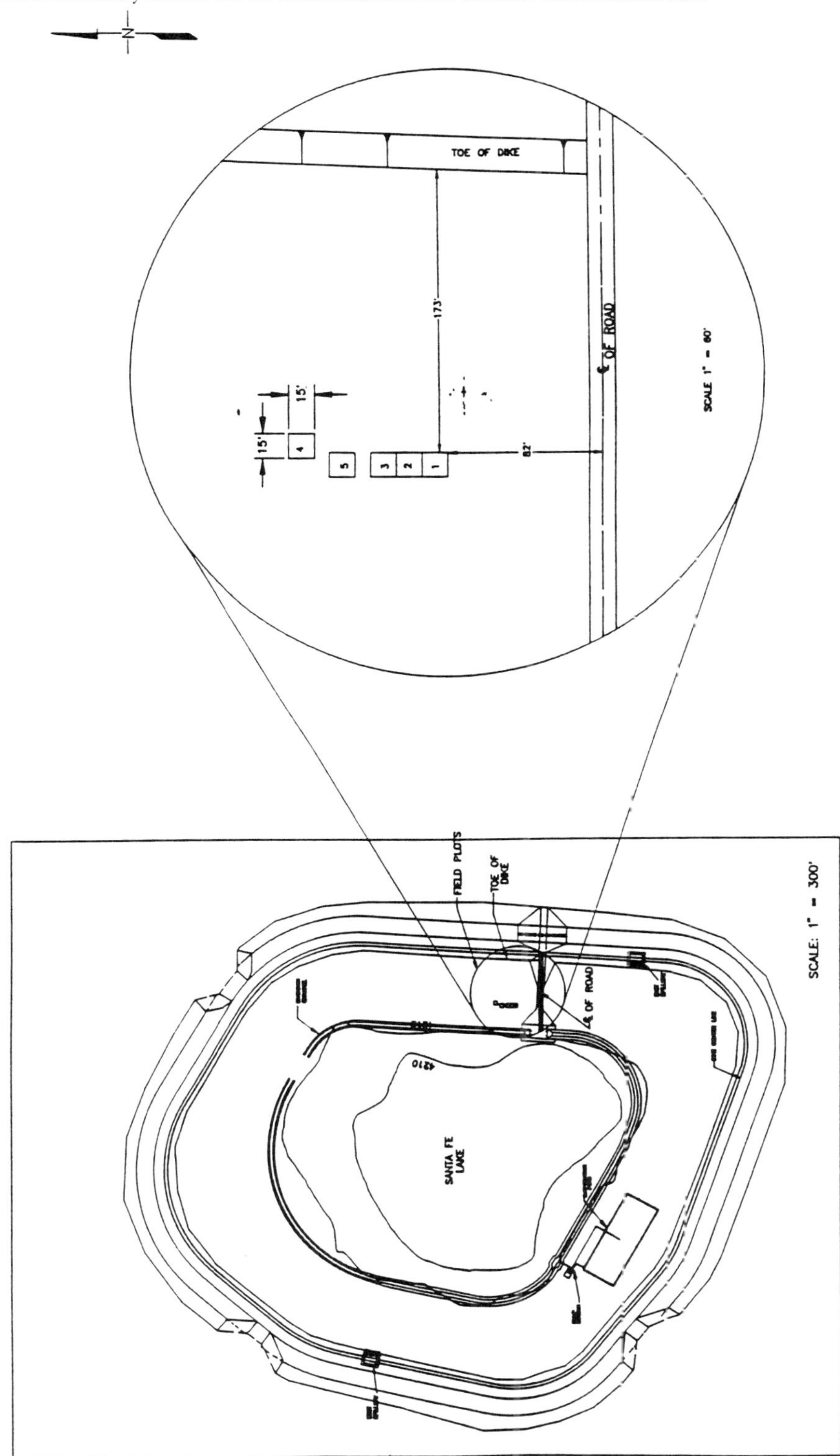

Figure 9-6.   Location and Layout of Field Plots.

Phase III includes the following, full-scale landfarming activity and the construction of an on-site storage facility. Phase IV will be the restoration of the Santa Fe Lake site. Figure 9-7 illustrates all systems directly and indirectly supporting biodegradation efforts.

Phase I and Phase II have been completed and Phase III is well underway. The construction for the on-site storage facility has been completed and full-scale bioremediation of the beach soils has been in operation for greater than one year.

The Phase I run-on dike is an earthen dike that encircles the contaminated area within the lake site. The high water mark for the lake system marks the inside toe of the dike. The dike embankment is approximately 12 feet high and covers a linear distance of approximately 4,200 feet. All structural fill material was untainted and obtained from within the lake site area. A 40-mil thick (1.0mm) high density polyethylene (HDPE) liner extends from the crest of the dike to just beyond the outside toe of the dike. This liner was necessary because of the collapse potential of the dike soils; it also serves as an erosion deterrent for the run-on dike system. The dike is equipped with two concrete emergency spillways and a series of piezometers. This dike was constructed to contain run-on waters from major storm events, outside of the contaminated area. The primary goal for the run-on dike was to enhance natural evaporation of the lake by preventing additional waters from running into the playa lake.

Phase II consisted of the construction of an automated irrigation system and a storm water management system. The storm water management system consists of a storm water diversion channel and an evaporation pond. This system handles the volume of rainfall that falls inside of the run-on dike. The storm water channel is divided into two components. The east channel routes storm water falling on the eastern half of the contaminated area, whereas the west channel routes storm water falling on the western half of the contaminated area. These two channels are hydraulically connected and are sloped to drain from north to south. Once the two channels meet, the waters gather at the headwater pool which is the inlet to the wetwell pump station. The water gravity flows from the headwater pool area to the wetwell chamber no. 1 pump until the PUMP ON float has been tripped. This pump transports storm water to an evaporation pond with a capacity to contain $78,000\,ft^3$ (approximatly 10,425 gallons) of storm water. Once the storm water reaches the basin, the twelve nozzle sprinler system vertically propels the water into the air in a fine mist form. This mist coupled with the high evaporation rate for the Clovis, New Mexico area allows the storm water to evaporate at an accelerated rate. Droplets that fall into the pond are routed to wetwell chamber pump no. 2. Pumping continues to take place from chamber no. 1 until the level in chamber no. 2 reaches the PUMP ON position. Once pump no. 2 has been activated the pump recycles the storm water until the water level falls below that minimum PUMP ON level. Figure 9-8 outlines the mechanism for storm water collected in the wetwell pump evaporation system. Both the storm water channel and the evaporation pond are lined with a 30-mil thick (0.75mm) HDPE liner.

The spray irrigation system was constructed to provide moisture to the treatment zone during Phase III operations. This system consists of an on-site groundwater well capable of delivering 220 gallons per minute of groundwater to the beach treatment area. Twenty-five rotating sprinklers were strategically placed around the beach area.

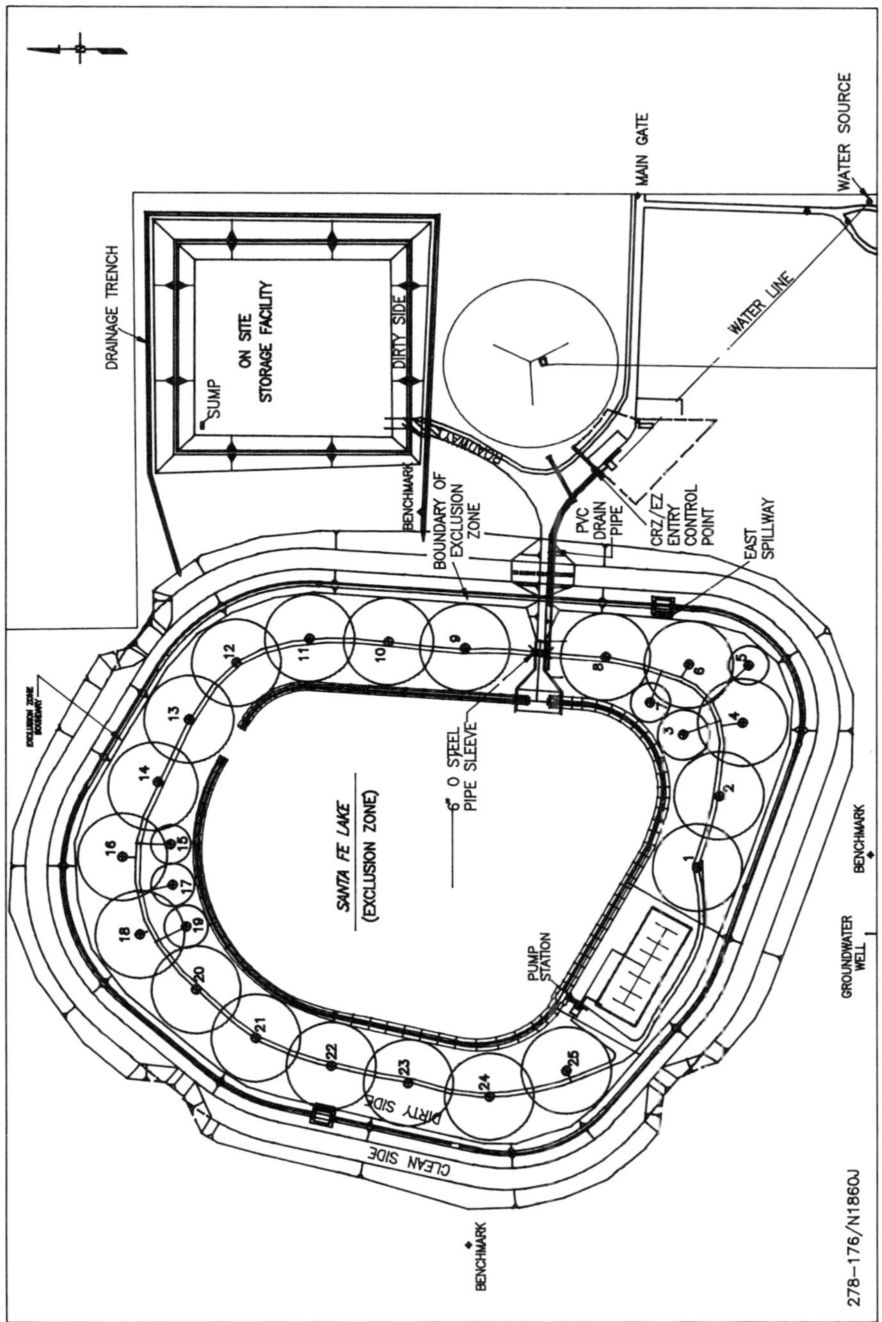

Figure 9-7. Sante Fe Lake Physical Works.

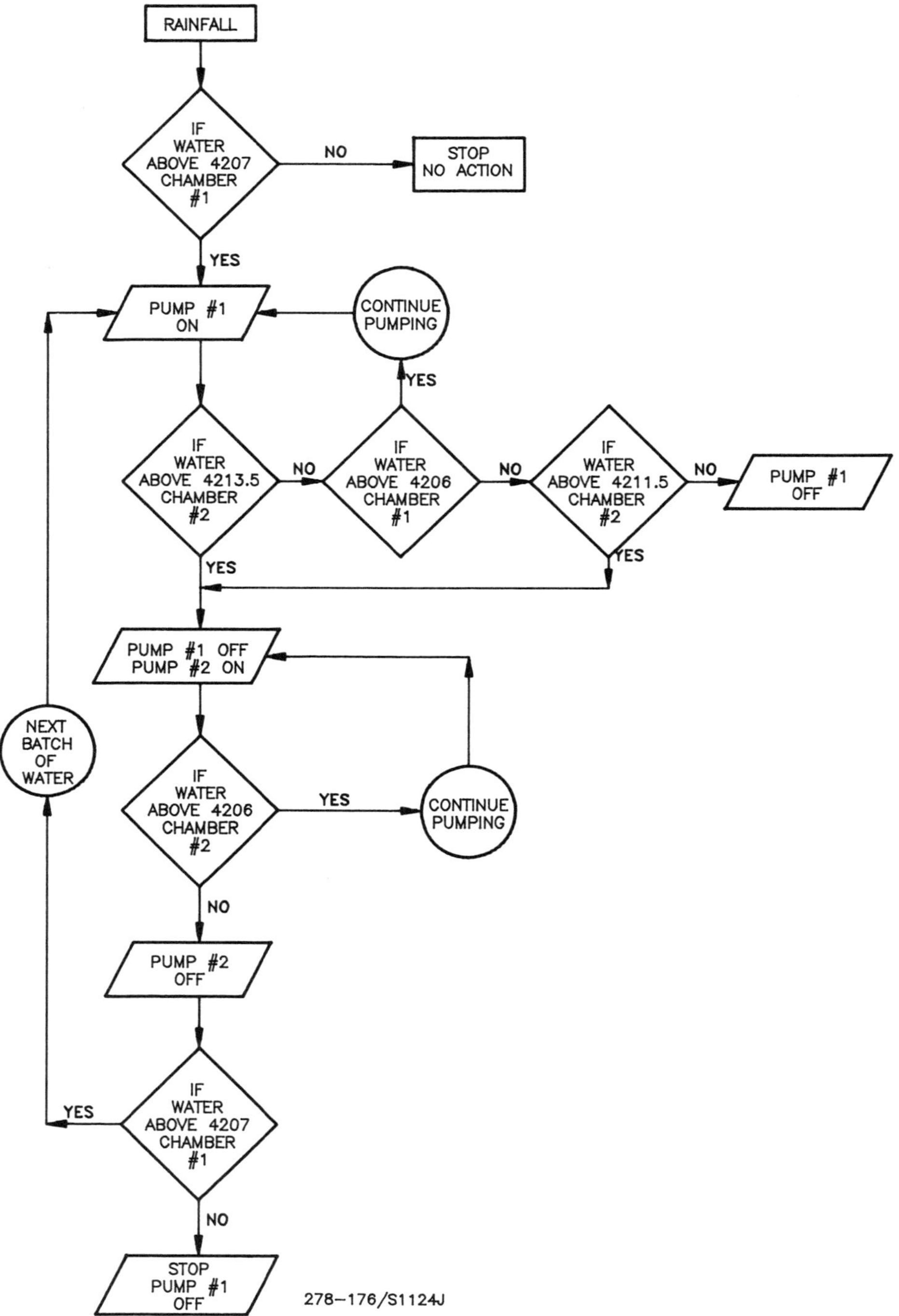

Figure 9-8. Flow Chart of Pump Station.

Each sprinkler is equipped with a tensiometer (irrometer). The irrometers sole purpose is to inform the system that irrigating water is or is not necessary. Irrigation takes place primarily during the night time hours when evaporation is at its lowest. However, seasonal changes require adjustments to be made to the irrigation schedule. This system is fully automated and once programmed needs only minor attention and that usually entails inputting a new irrigation schedule in or any necessary repair or maintenance work.

A decontamination station was constructed for use during Phase II, III, and IV. The decontamination station is equiped with a 2,000psi pressure washer to accomodate decontamination of equipment and site personnel. Rinsate collected from equipment washing and personnel decontamination are routed via pipeline to the east storm water diversion channel.

Phase III consisted of the construction of an on-site storage facility. All treated contaminated soils stabilizing above the 1,000 ppm cleanup level must be transported to the OSF; all treated soils that fall below the cleanup level may be left in place. All lake bottom sediments are to be transported to the OSF for treatment. The on-site storage facility is a 4.5 acre treatment cell equipped with 11-foot earthen dikes. This system is responsible for treating all sediments via the same basic landfarming technique incorporated in the beach area. The OSF also serves as a final resting place for soils stabilizing above 1,000 ppm and any contaminated construction debris. This system is capable of housing all presently contaminated beach soils and lake-bottom sediments. It is also equipped with a small storm water management system that transports the rainfall volume that impacts the area within the OSF. This storm water management system was built to tie into the decontamination conveyance system. This system is also fully automated.

A small irrigation system will be placed within the OSF to provide moisture for the biodegradation process prior to the placement of a lift of material within the OSF.

Phase IV will consist of the capping of the OSF with a PVC or other pre-approved synthetic liner and signifies the restoration of the site. After capping ends, the dikes will be demolished and all ditches filled in. The site will be regraded and revegetated.

FULL-SCALE BIOREMEDIATION

In July 1992, full-scale bioremediation was initiated at the Santa Fe Lake site. Full-scale bioremediation required the use of a 140 horsepower (HP) tractor, a disk implement with 26-inch diameter disks, a till-oll implement, a 5-ton fertilizer spreader, and a frontend loader attachment. The landfarming operation took place much the same way that the field plot study was conducted. The Santa Fe Lake site operator cultivates the soil one week with the disk implement and the next week the till-oll implement is used. Those particular implements are responsible for aerating the soil and for maintaining minimum clod sizes. The finer the soil has been churned up, generally the better biodegradation proceeds. Ideally, the disk implement is responsible for "flip-flopping" the top 12 inches of soil. In essence, this implement places material that currently rests at the bottom of the 12 inches at the top surface of the twelve inches and vice versa. The till-oll implement primarily affects the top six inches. This implement is equipped with spherical disks, shovels, and a rotating element that thoroughly pulverizes soil clods leaving the tilled area fluffy and homogenized. This action increases void sizes within the soil thus improving the oxygen transfer from the atmosphere into the treatment zone. The fertilizer spreader distributes ammonium sulfate fertilizer to the beach soils on a monthly basis. Generally, the beach soils are irrigated from Tuesday through Saturday.

The beach soils have been divided into eight sections for sampling identification purposes. Samples are taken at the beginning of the lift cultivation period and at the end of two months. If samples taken meet the 1,000 ppm cleanup level, then cultivation in that section may cease. All sections that do not reach the 1,000 ppm cleanup level must achieve a stabilization criteria. Stabilization can be achieved when two out of three consecutive sampling events indicate that no further significant biodegradation has taken place during this time period.

Presently, biodegradation for the first lift has gone well and as soon as the sediments can be placed within the OSF, then lift no. 2 for beach soils will get underway. Lake bottom sediment biodegidation will commence once the OSF irrigation system has been installed and is operational.

A rigorous sampling plan has been established and followed to assess the biodegradation process and how it affects not only the treated soils but the soil layers underlying the treatment zone and the groundwater.

In conclusion, biodegradation of soils and sediments at the Santa Fe Lake site has been extremely successful. Due to careful planning and successful experimentation with the site parameters, a mechanisim has been implemented that allows for efficient degradation of the site contaminants.

Questions and Answers:

Q. Do you have any oil field waste there?

A.   No, there's no oil field waste.   The railyard itself was the only source of hydrocarbons.

Q. Did you do any testing for radiological material?

A. Radiological? No we did not. Typically, at railyards and railroad facilities, radiation is not a problem and it was not considered.

Q. Did I hear you say that the regulatory people limited you to twelve inch lifts.

A. I wouldn't say that they mandated that. That was in our work plan that 12 inches was an appropriate lift depth and it was accepted by the EPA.

Q. Was there some reason that you didn't try to go to a deeper lift depth?

A. Typically at other sites we've gone to 18 inch lifts but that was more porous soil, and this was a denser soil. It had a lot of biomass in it, sediment left over from the hopper car washing facility. And being a denser material, we thought that the shallower lift would be more appropriate for aerating that lift.

Q. You talked about that thousand parts per million (ppm) goal that EPA set for you and discussed in brief the leveling off of the biological activity. How do you determine that the bacteria has leveled off it's activity and are you certain that in the future it won't recharge through desorption episodes.

A. I've seen this stabilizing before. I understand it's quite common at sites where you'll go from 10,000 ppm down to a 1,000 ppm and stabilize, and then take a second step and possibly go down from 1,000 ppm to 100 ppm. That was a consideration here, but we were able to reach an agreement with EPA. I can't say whether those microbes will kick in again at this site or not. We conducted a risk assessment that showed that 10,000 ppm is an acceptable cleanup level. EPA was not comfortable with that and proposed a level of 1,000 ppm which we accepted.

Q. And that risk assessment was based on land use?

A. Yes, ultimate land use as well as other considerations as outlined in the California LUFT manual such as depth to groundwater, types of soils, and other miscellaneous categories.

CHAPTER 10

Results of a Pilot Study for Aboveground Bioremediation of Number 2 Diesel Fuel Contaminated Soils

Peter R. Guest, P.E., Kent A. Friesen, P.E., and E. Kinzie Gordon
Engineering-Science, Inc. — Denver, Colorado

INTRODUCTION

A pilot test was performed using enhanced aboveground biological treatment to remediate petroleum-contaminated soils excavated during an underground storage tank (UST) removal project at Lowry Air Force Base (AFB), Colorado. Following confirmation of a release, over 5,000 cubic yards (yd^3) of soil contaminated with No. 2 fuel oil was removed from the tank excavation in April 1992, and placed into a full-scale biotreatment facility at Lowry AFB. The purpose of the pilot test was to determine if the addition of nutrients and/or commercially available bacteria is a cost-effective means of enhancing the full-scale biological treatment of the fuel-contaminated soils. The pilot test was conducted concurrently with full-scale treatment at the facility. A summary of the pilot test methodology, results, and conclusions is presented in this chapter.

BIODEGRADABILITY OF HYDROCARBONS

Degradation of fuel hydrocarbons in soils by bacteria and other microorganisms has been used extensively as a remediation technique and is well documented in the technical literature. Hydrocarbon degrading bacteria have been found in almost all soil environments, and typically occur in high numbers when petroleum is present.[1] Many successful field demonstrations of enhanced biodegradation of gasoline, diesel, and jet fuels in soil using *in situ* and aboveground methods have been reported. Aboveground bioremediation utilizes the growth of naturally occurring microorganisms in the soil to biodegrade petroleum hydrocarbons in the presence of an adequate supply of oxygen, moisture, and nutrients.

Hydrocarbon degrading soil bacteria require sufficient oxygen, nutrients, and moisture to attain the activity levels required for effective soil treatment. Because insufficient oxygen is often the primary limiting factor for petroleum biodegradation, aeration of contaminated soil is performed to provide sufficient oxygen for soil bacteria.

This is generally accomplished through soil spreading and tilling (i.e., landfarming), which supplies oxygen to the soil microorganisms and produces a homogeneous mixture of soil. Periodic moisture and nutrient addition may also be required. Once sufficient oxygen, nutrients, and moisture are provided, natural bacteria in the soil multiply and thrive using fuel hydrocarbons as their primary carbon source.[2] Optimal temperatures for petroleum biodegradation have been reported as 68 degrees Fahrenheit (°F) to 104°F. Biodegradation can continue at temperatures above freezing; however, the rates of degradation can decrease significantly with decreasing temperatures.[2]

Another factor influencing hydrocarbon biodegradation is the availability of nutrients, such as nitrogen and phosphorus, to soil microbes. A carbon-to-nitrogen (C:N) ratio of 20:1 is considered optimum for fuel biodegradation in soils.[3] Lack of adequate phosphorus may also limit biodegradation. Research on jet fuel contamination at Tyndall AFB suggests that soil bacteria recycle nutrients, and may rely on nitrogenous bacteria to fix atmospheric nitrogen, thereby introducing useful forms of nitrogen for fuel-degrading microbes.[4]

Adequate soil moisture is also required to sustain growth of hydrocarbon degrading microbes. However, excessive saturation of soil with water will reduce air permeability in soil, result in a lower oxygen supply, and consequently inhibit biodegradation. Saturation between 50% and 80% of the water holding capacity of a soil is considered optimal for aerobic microbial activities.[2]

Aboveground treatment of hydrocarbon contaminated soils has several advantages over *in situ* treatment. Landfarming of excavated soils produces a more homogeneous mixture of sands, silts, and clays. In the *in situ* environment, sand and clay lenses frequently result in a nonuniform distribution of fuels, making uniform treatment more difficult. Oxygen, nutrients, and moisture can be evenly distributed in aboveground operations, and hydrocarbon treatment is more rapid and uniform. Finally, the sampling of soils in aboveground operations is comparatively simple and more reliable than sampling in soil borings, which are subject to subsurface nonuniformities.

Numerous studies have demonstrated petroleum degradation in soils by indigenous microbes.[1,5,6] However, currently there are many companies which claim to have isolated natural bacteria with unique capabilities for increasing the rate of natural fuel biodegradation. The scientific literature is generally not conclusive regarding the viability of inoculated soil remediation (also known as bioaugmentation) when compared to the rate of degradation produced by providing an increased oxygen and nutrient supply to indigenous soil bacteria (also known as biostimulation). However, low degradation rates have been reported for soils recently contaminated by a petroleum product release, possibly because indigenous bacteria have not become acclimated to the sudden change in soil condition[7]. This phenomenon suggests that an evaluation of the potential benefit of innoculation of recently contaminated soils with vendor-supplied bacteria is justified.

PILOT TEST METHODS

The pilot test was performed over a 6-month period from July through December 1992. Six wooden boxes with capacities of 2.5 yd^3 each were constructed as biotreatment cells for the test (Cells 1 through 6). Each cell represented a discrete combination of soil amendments and maintenance activities, with the exception of Cell 6 which served as a control. The effects of the various amendments on degradation of fuel in soils were monitored through periodic sampling and analysis of test cell soils.

Soil Amendments and Cell Maintenance

Contaminated soils for the pilot test were obtained from the site during the UST removal activity. The soil was homogenized by mixing with a front-end loader, and 2.5 yd^3 of soil was placed into each of the six test cells. A description of the amendments applied to soils in each test cell is provided in Table 10-1. Three cells (Cells 1 through 3) were amended with commercial mixtures of bacteria and nutrients from three different vendors of bioremediation products. Vendor instructions were followed with regard to application rates and maintenance activities. The soil in Cell 4 was amended with ammonium nitrate fertilizer to assess biostimulation of indigenous bacteria by nutrient addition. Cell 5 soil contained no amendments to assess the effects of soil aeration only.

Weekly maintenance of the test cells included addition of water as needed and tilling of the soil in Cells 1 through 5. Based on field observations and monthly analysis of the soils for percent moisture, moisture content in the cells was adjusted by spraying water at a controlled rate on each cell prior to rototilling. Soil moisture was generally maintained above 10% (wet weight basis) to provide adequate water for biodegradation. Water addition (other than precipitation) and tilling were not performed on Cell 6, the control. Rototilling was performed by placing a 5-horsepower front-tine rototiller into the test cell and thoroughly mixing the soil. Detailed descriptions of the amendments and additional maintenance for each test cell are provided below.

Cell 1: Vendor A
Vendor A distributes a product containing live cultures of petroleum degrading facultative anaerobic bacteria suspended in a liquid medium, and a supplemental

Table 10-1. Test cell amendments

Test Cell	Description
Cell 1 - Vendor A	Bacteria in liquid medium and vendor-supplied nutrients
Cell 2 - Vendor B	Freeze-dried bacteria and ammonium nitrate fertilizer
Cell 3 - Vendor C	Incubated bacteria and nutrient mixture
Cell 4 - Nutrients	Ammonium nitrate fertilizer
Cell 5 - Till Only	Tilling only, no amendments
Cell 6 - No Till	No tilling, no amendments

nutrient product that consists of a proprietary blend of inorganic nitrogen and phosphorus compounds. In accordance with vendor directions, the nutrient solution was reapplied to the soils of Cell 1 57 days and 114 days after the initial application. The Vendor A bacteria were reapplied after 57 days.

Cell 2: Vendor B
Vendor B bacteria product, produced for enhanced biodegradation of heavier fuel products such as diesel or fuel oil, was applied to Cell 2. This product is a dry solid consisting of freeze-dried bacteria adhered to bran. The product is soaked in water prior to application to activate the bacteria. The Vendor B product was applied to the soil following application of a water-soluble ammonium nitrate fertilizer. This same nutrient amendment was used for Cell 4 (nutrient only). In accordance with vendor instructions, the Vendor B bacteria product was reapplied to Cell 2 soils 2 weeks, and then every 4 weeks, after the initial application.

Cell 3: Vendor C
Vendor C's nutrient and bacteria product, intended for degradation of no. 2 fuel oil, is freeze-dried and sold in a blend that contains an initial carbon source for activating the culture, as well as nitrogen, potassium, phosphorus, and trace minerals as nutrients. Because the product contains nutrients and bacteria, dual application of materials was not required. Prior to application, the Vendor C product requires incubation, involving mixing with water and aerating continuously for 24 hours. Properly incubated product was reapplied to the soils of Cell 3 once per week prior to tilling. The manufacturer recommended that the reapplication and tilling occur two or three times a week; however, the reapplication was limited to once a week to limit the advantage that increased tilling would provide in relation to the other test cells.

Cell 4: Nutrients
Agricultural ammonium nitrate fertilizer was added to Cell 4 to determine the rate of hydrocarbon degradation attributable to the indigenous bacteria when supplemented with additional nitrogen. The initial application of fertilizer was designed to provide a C:N ratio of approximately 60:1, which is roughly three times the amount considered optimum for biodegradation in literature references.[3] These nutrients, above optimum levels, were provided to ensure that the absence of nutrients would not be limiting.

Cell 5: Till Only
As oxygen depletion is commonly a limiting factor in hydrocarbon degradation, Cell 5 was maintained to determine the effectiveness of hydrocarbon degradation by indigenous bacteria given ample oxygen but no additional nutrients. No soil amendment other than water was applied to this test cell. No maintenance other than weekly tilling and moisture addition (as needed) was performed.

Cell 6: Control
The control cell served to quantify the degradation of fuel in soil that would occur if no amendments were added and no maintenance was performed. After placement of soils in Cell 6, no activities other than sampling were performed. Moisture (other than precipitation) was not added to Cell 6.

Sampling and Analysis

Data for the pilot test were based on initial, periodic, and final sampling and analysis of soils from the six test cells. Soil samples collected from the test cells were composite samples collected from six locations within each cell. The initial sampling for total petroleum hydrocarbons (TPH) and benzene, toluene, ethylbenzene, and xylenes (BTEX) was performed prior to the initial application of nutrients and bacteria to the test cell soils. Following application of nutrients and bacteria products, the soil from each of the six test cells was sampled for nutrient and bacteria analysis. Periodic sampling was performed approximately monthly for six months. After the initial sampling event, soil sampling was performed prior to rototilling and application of water or reapplication of amendments. Quality assurance samples collected during the pilot test included field duplicate samples from one test cell during each of the sampling events.

Soil samples were analyzed for the parameters listed in Table 10-2. Analytes included TPH and BTEX for characterization of petroleum constituents; nitrogen, phosphorus, and pH for characterization of nutrient requirements; percent moisture to monitor soil moisture; and bacteria assay. The bacteria assays were performed by the Agronomy Department at Colorado State University (CSU), Fort Collins, Colorado. CSU has developed techniques for determination of heterotrophic bacteria populations, and hydrocarbon degrading bacteria populations utilizing plate-count methods.

Table 10-2. Soil analytical methods

Parameter	Method
TPH (Diesel)	SW8015 modified
BTEX	
Benzene	SW 5030/8020
Ethylbenzene	SW 5030/8020
Toluene	SW 5030/8020
Xylenes	SW 5030/8020
Nitrogen	
Total Kjeldahl Nitrogen	E351.2
Ammonia	E350.3
Nitrate-Nitrite	E353.2
Water-Soluble Phosphorus	E365.4
pH	150.1 modified
Percent Moisture	3500
Heterotrophic Bacteria	CSU[a] technique
Hydrocarbon Degrading Bacteria	CSU technique

[a] CSU=Colorado State University

PILOT TEST RESULTS

The labor and logistical considerations involved with the application of the different amendments varied significantly with each amendment. Because Vendor A's bacteria product consisted of bacteria suspended in a liquid medium, the application of this product to Cell 1 proved to be the least labor intensive. Vendor A indicated that two applications of the bacteria product were adequate for the duration of the pilot test. Vendor B's bacteria product required soaking the dry bran product for 4 to 6 hours prior to filtering and spraying the inoculate onto the soil cells. The Vendor B product was reapplied monthly. Vendor C's bacteria product generally required the greatest amount of preparation prior to application, as the culture was incubated overnight prior to application. Vendor C indicated that application of the inoculate should be performed at least weekly for best results.

The pilot test was conducted from July through December. Freezing temperatures in November prevented adequate tilling of the cells due to frozen soil. Larger and more frequent water additions were required during the hot and dry summer and fall months to maintain 10% moisture in the soil.

Analytical results for the landfarm pilot test soil samples are presented in Table 10-3, which is a compilation of the pilot test data organized to delineate trends in the analyses over time. In Table 10-3, the analytical results for each test cell have been grouped, and the results for field duplicates have been averaged. The pilot test results for each analytical method or group of methods are discussed below.

Total Petroleum Hydrocarbon Results

The TPH results (dry-weight basis) are presented in Figure 10-1, which presents TPH concentrations in soils with respect to time for each of the six test cells. Initial TPH concentrations ranged from 3,000 milligrams per kilogram (mg/kg) in Cell 1, to 5,000 mg/kg in Cell 2. The average initial TPH concentration in all six cells was 3,800 mg/kg.

Cells 1 through 5 exhibited an overall trend of decreasing TPH concentrations during the 6-month pilot test. Final TPH concentrations in the cells with vendor bacteria added were 800 mg/kg for Cell 1 (Vendor A), 800 mg/kg for Cell 2 (Vendor B), and 1,200 mg/kg for Cell 3 (Vendor C). Final TPH concentrations were 1,600 mg/kg for Cell 4 (nutrients), and 1,900 mg/kg for Cell 5 (till only). The Cell 6 control samples exhibited an overall increase in TPH concentrations (from 4,000 to 5,300 mg/kg).

TPH data are somewhat irregular for all of the cells sampled. Although a consistently decreasing (or stable) trend would be expected due to degradation (or lack of degradation), instead there is an overall decreasing trend in Cells 1 through 5 with a significant increase in TPH concentrations measured during the fourth sampling event. The irregularity in TPH results is believed to be related to the potential for stratification of TPH in the soil cells, and to the field sampling and analytical variability that is inherent in soil sampling. The degree of moisture and petroleum stratification was observed to be greatest following periods of heavy precipitation, which could cause leaching within the soil cells. In addition, Cell 6, which exhibited the greatest amount of irregularity in TPH results, was the one cell that was never tilled, thereby eliminating the mixing that would diminish the problem of TPH stratification in soils.

Even with the irregularities in TPH results, an overall decreasing TPH trend that can be attributed to biodegradation was exhibited in Cells 1 through 5 (Figure 10-1). The analytical results for Cell 4 (nutrients added) as compared with results for Cells 1 through 3 (vendor bacteria and nutrients) demonstrate that the use of commercially available bacteria cultures produced greater TPH degradation rates than indigenous bacteria alone, as calculated from initial and final TPH concentrations in soil from each

Table 10-3. Summary of pilot test soil data by cell

	Day	TPH (wet wt.) (mg/kg)	Moisture (%)	TPH (dry wt.) (mg/kg)	Hydrocarbon Degraders (CFU/g or ml)	Total Heterotrophic Bacteria (CFU/g or ml)
Cell 1	0	2700	8.8	3000	593000	6010000
	28	1500	12	1700	25200000	26750000
	56	900	11	1000	26900000	89400000
	84	1800	5.4	1900	43600000	84700000
	112	1200	6	1300	30500000	97100000
	161	710	13	800	81300000	171000000
Cell 2	0	4600	7.2	5000	434000	12300000
	28	1900	11	2100	21700000	26000000
	56	1800	11	2000	20450000	51300000
	84	2400	6.9	2600	28400000	98100000
	112	1720	6	1800	21600000	100000000
	161	650	19	800	65100000	236000000
Cell 3	0	3900	6.1	4200	7860000	30600000
	28	2100	11	2400	27200000	66500000
	56	1400	10	1600	20300000	36800000
	84	2300	7.1	2500	15050000	56050000
	112	1200	8	1300	12100000	52100000
	161	1000	15	1200	100000000	157000000
Cell 4	0	3000	6.5	3200	856000	10500000
	28	1500	11	1700	23100000	66200000
	56	1300	11	1500	34300000	90800000
	84	3300	6.1	3500	29100000	64600000
	112	2000	8	2200	38900000	86500000
	161	1400	14	1600	62100000	114000000
Cell 5	0	3200	6.9	3400	10700000	17500000
	28	1900	11	2100	24900000	54800000
	56	1800	11	2000	450000	19300000
	84	2700	6.5	2900	3300000	31000000
	112	2600	9	2900	6350000	15100000
	161	1550	17.5	1900	11150000	22900000
Cell 6	0	3750	7	4000	9765000	45000000
	28	1600	7	1700	14900000	46000000
	56	2500	10	2800	3000000	24400000
	84	6500	7.8	7000	15100000	36600000
	112	4900	6	5200	5220000	16100000
	161	4600	13	5300	2550000	27100000

cell. The cells with Vendor A bacteria and nutrient products (Cell 1) and Vendor B bacteria product and fertilizer (Cell 2) showed the greatest decrease in TPH concentrations. Clearly, the least amount of TPH biodegradation occurred in Cell 6, which exhibited an apparent increase in TPH concentrations.

Table 10-3.   (Cont.)

pH	TKN (mg/kg)	Nitrite-Nitrate (mg/kg)	Total Nitrogen [a] (mg/kg)	Ammonia (mg/kg)	Total Phosphate (mg/kg)	Total BTEX (μg/kg)
8.07	90	1	91	2	7	51
8.05	135	1	136	3	8	7.8
8.00	100	1	101	4	27	NA [b]
8.20	150	1	151	3	21	NA
7.10	280	1	281	2	29	NA
7.90	170	1	171	28	37	NA
8.45	410	140	550	160	8	292
7.90	340	55	395	81	4	10.7
8.45	175	2	177	45	5	NA
8.60	190	4	194	47	5	NA
8.80	320	5	325	44	2.3	NA
8.50	200	1	201	44	12	NA
8.70	76	1	77	4	15	282
8.00	110	1	111	6	39	21.3
8.20	67	1	68	4	130	NA
7.90	69	1	70	4	230	NA
7.60	210	1	211	3	430	NA
7.70	160	1	161	14	500	NA
8.48	360	180	540	120	4	245
7.70	290	120	410	110	4	6.2
8.20	240	40	280	70	3	NA
8.30	210	46	256	61	4	NA
8.25	330	44	374	75	10	NA
8.40	220	6	226	57	12	NA
8.95	67	1	68	4	6	504
8.40	100	1	101	6	6	12.8
8.40	25	1	26	3	1	NA
8.50	25	1	26	3	7	NA
8.90	92	1	93	4	57	NA
8.95	52	1	53	1	12	NA
8.97	75	1	76	2	4	344
8.25	82	1	83	2	3	9.2
8.50	25	1	26	1	5	NA
8.30	25	1	26	2	1	NA
8.60	70	1	71	4	3	NA
9.00	56	1	57	3	5.9	NA

[a] Total nitrogen equals the sum of the total Kjeldahl nitrogen (TKN) and nitrate-nitrite as nitrogen.
[b] NA = not analyzed.

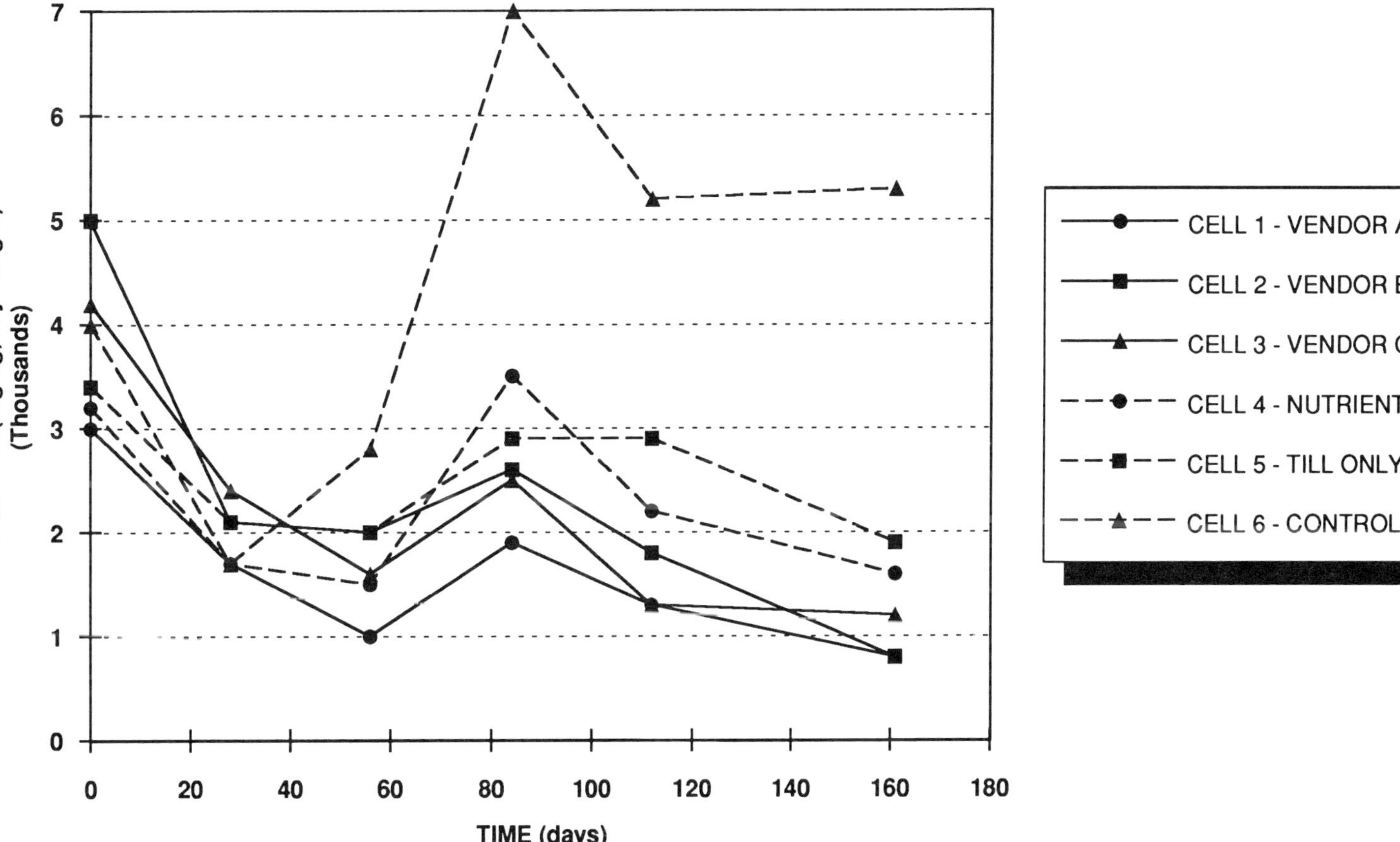

Figure 10-1.　Total petroleum hydrocarbon pilot test results

Evaluation of TPH Chromatograms

Chromatograms from the initial (day 0) and final (day 161) TPH analyses were compared for each cell to qualitatively assess the constituents that were removed during the pilot test. The first and last sample chromatograms for Vendor A (Cell 1) and till only (Cell 5) are presented in Figures 10-2 and 10-3, respectively. Qualitative identification of the peaks in the TPH chromatograms from the pilot test soil samples is shown.

It is apparent that hydrocarbons in the range of C11 to C24 were originally present in the fuel oil released to the soil. The larger peaks lie within the range of C14 to C18 hydrocarbons. The prominent peak at C26 is a surrogate compound. There are doublet peaks at C17 and C18 that probably correspond to C17/pristane and C18/phytane. Pristane and phytane are isoprenoid compounds that are present in petroleum, but are generally fairly resistant to biodegradation, according to the literature.[8,9] Although positive identification of the isoprenoid compounds cannot be made with the analyses performed, the peak patterns of the chromatographs lend themselves well to qualitative matching of the C17/pristane and the C18/phytane doublet peaks.

Final soil samples from Cell 1 contain hydrocarbon peaks in the range of C14 to C22, with the larger peaks within the range of C16 to C18. A significant number of peaks in the range of the C11 to C16 hydrocarbons have been removed in comparison with the chromatogram from the initial soil sample, presumably by biodegradation. Similar observations were made with regard to Cell 2, 3, and 4 initial and final chromatograms, suggesting that biodegradation was stimulated by the addition of nutrients and bacteria to the soil plots. Biodegradation was also suggested by the qualitatively identified decrease in the C17/pristane and C18/phytane ratios in Cells 1 and 3 (Figure 10-2). Chromatograms from Cells 5 and 6 generally contain similar peak patterns in the initial soil samples and final soil samples (Figure 10-3).

The chromatogram observations suggest that preferential removal of certain compounds over others with similar molecular weights is occurring. This result is consistent with previous observations of preferential biodegradation of some compounds (such as straight alkanes and aromatics) over other compounds (such as branched alkanes and alkyl-substituted aromatics) that have been documented in the literature.[10] Generally, it was also observed that the chromatograms from the final soil samples contained more unresolved components and fewer real peaks than the initial soil sample chromatograms. It was proposed that some of the initial hydrocarbons may have been transformed into different compounds that chromatograph poorly (i.e., polar compounds such as carboxylic acids).

It is interesting to note that the unresolved peaks were strongly developed in the final chromatogram of Cell 6 when compared to the initial chromatogram. It is possible that the increasing TPH concentrations measured over time in Cell 6 are actually a relic of chemical transformation (e.g., weathering) of the fuel oil rather than a result of field sampling variability.[8]

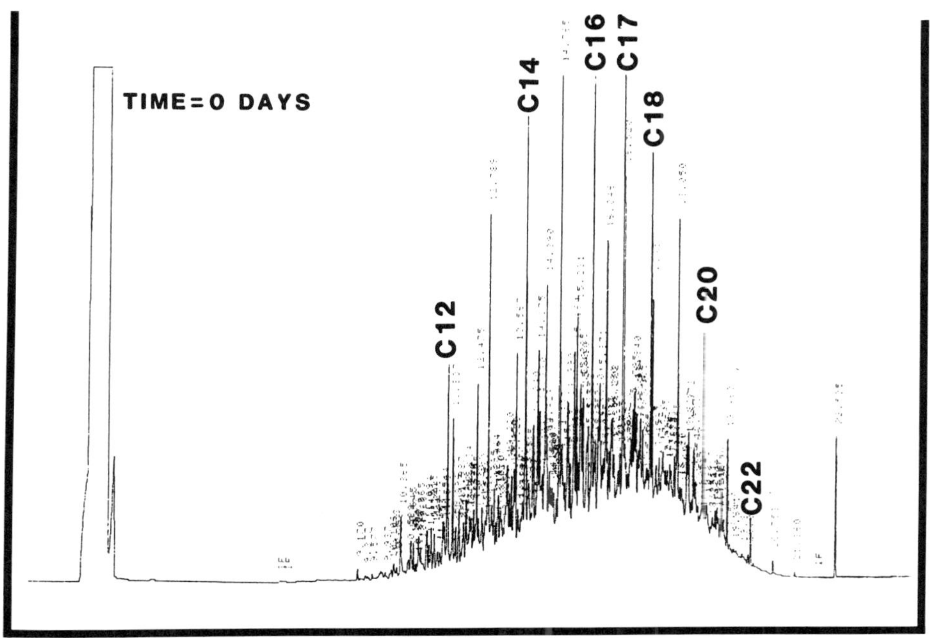

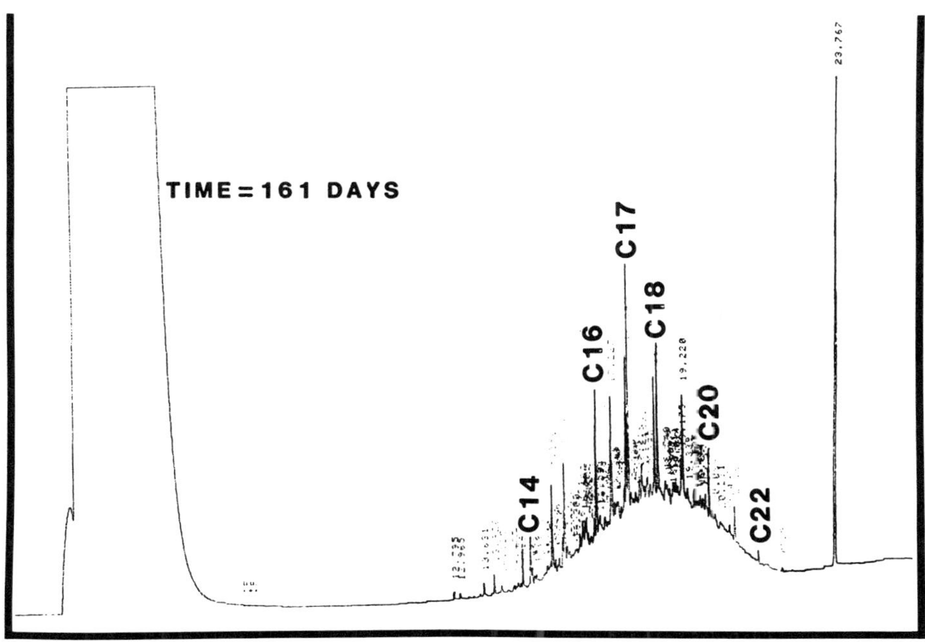

Figure 10-2. Representative soil TPH chromatograms showing hydrocarbon peaks for Cell 1 (Vendor A) during pilot testing

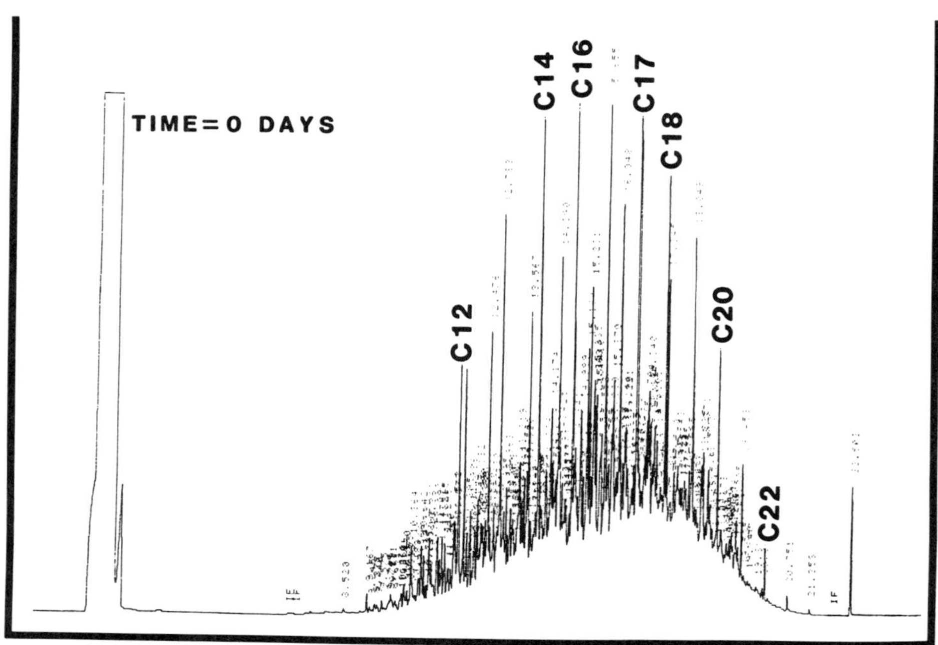

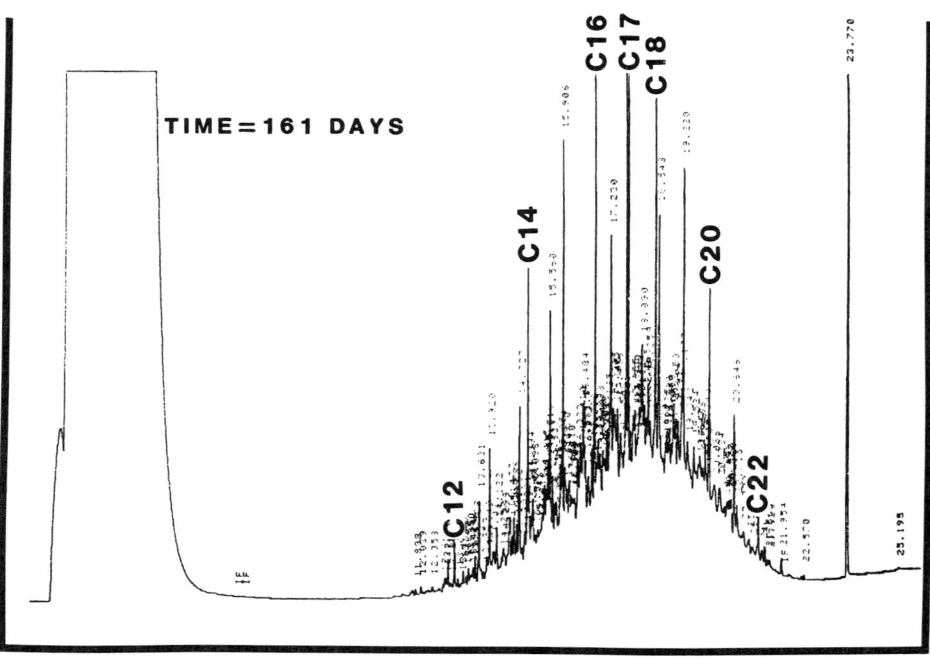

Figure 10-3. Representative soil TPH chromatograms showing hydrocarbon peaks for Cell 5 (till only) during pilot testing.

Bacteria Assay Results

Samples of the test cell soils and the initial bacteria inoculants were collected for determination of total heterotrophic bacteria and hydrocarbon degrading bacteria. The initial inoculants all contained at least 200,000 colony forming units per milliliter (CFU/ml) of inoculant of hydrocarbon degrading bacteria. The inoculants from Vendor A and Vendor B contained at least two orders of magnitude greater quantities of microbial cells than the Vendor C product. All three inoculants contained a large number (11 million to 178 million CFU/ml) of heterotrophic bacteria.

All of the test cells, including the control cell, contained a large quantity of heterotrophic bacteria throughout the duration of the pilot test. A graph of the heterotrophic bacteria populations in the cells over time is presented in Figure 10-4. This figure shows a general trend of increasing heterotrophic bacteria populations with time in Cells 1 through 4, which include the cells with vendor bacteria additions as well as the cell with nutrients only added. Cells 5 (till only) and 6 (control) show a stable to decreasing trend of heterotrophic bacteria populations with time.

It is apparent that the addition of nutrients and/or bacteria inoculants led to an increase of approximately an order of magnitude in bacteria populations for Cells 1 through 4 over time. The addition of nutrients resulted in bacteria counts comparable to those in the cells with addition of nutrients and commercial bacteria cultures. A slight decrease in the population of heterotrophic bacteria is apparent in Cell 5 (till only) and Cell 6 (control).

Graphs of the hydrocarbon degrading bacteria populations in the cells over time are presented in Figure 10-5. Similar to the heterotrophic bacteria data presented in Figure 10-4, this figure shows a general trend of increasing hydrocarbon degrader populations with time in Cells 1 through 4, which includes the cells with vendor bacteria additions as well as the cell with nutrients only added. Cells 5 (till only) and 6 (control) show a stable to decreasing trend of hydrocarbon degradering bacteria populations over time. The addition of nutrients and/or bacteria cultures led to an increase of two orders of magnitude for Cells 1 and 2 over time, and one order of magnitude for Cells 3 and 4 over time.

BTEX Results

Total BTEX concentrations of 0.051 mg/kg to 0.504 mg/kg were measured in the initial soil samples from the test cells (Table 10-3). Since these initial concentrations were already well below the most stringent state soil cleanup guidelines for total BTEX (20 mg/kg), the BTEX monitoring was discontinued following the second sampling event. Significant reduction in total BTEX was shown in all of the test cells, including the Cell 6 control, following 28 days of treatment. Total BTEX concentrations after 28 days ranged from 0.0062 mg/kg to 0.0128 mg/kg.

Nitrogen Results

The total Kjeldahl nitrogen, ammonia, and nitrite-nitrate analyses results provided in Table 10-3, indicate that it is likely that the nutrient products supplied by Vendor A (Cell 1) and Vendor C (Cell 3) contained little nitrogen in inorganic forms such as nitrate or ammonium compounds. Nitrite-nitrate and ammonia concentrations were elevated in Cells 2 and 4, as would be expected based on the addition of ammonium

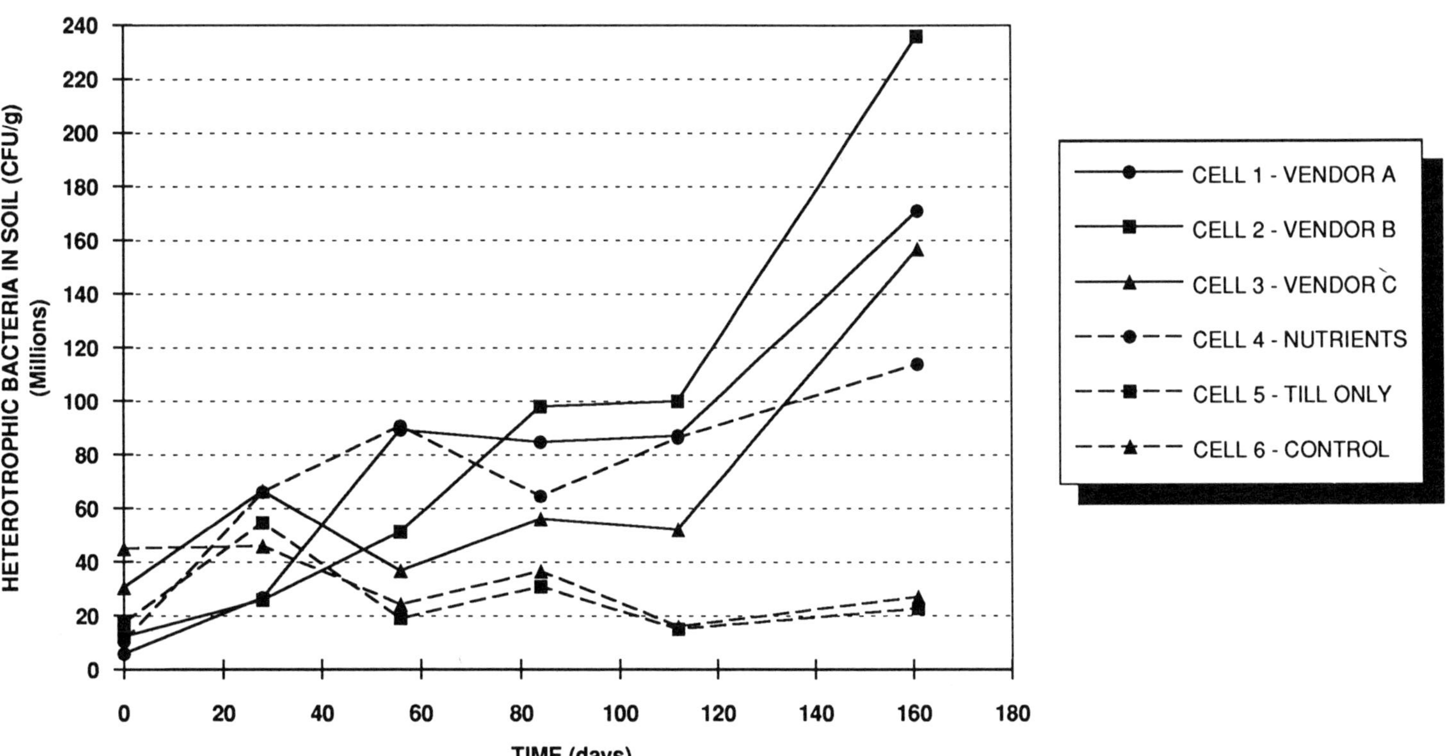

Figure 10-4.   Total heterotrophic bacteria pilot test results

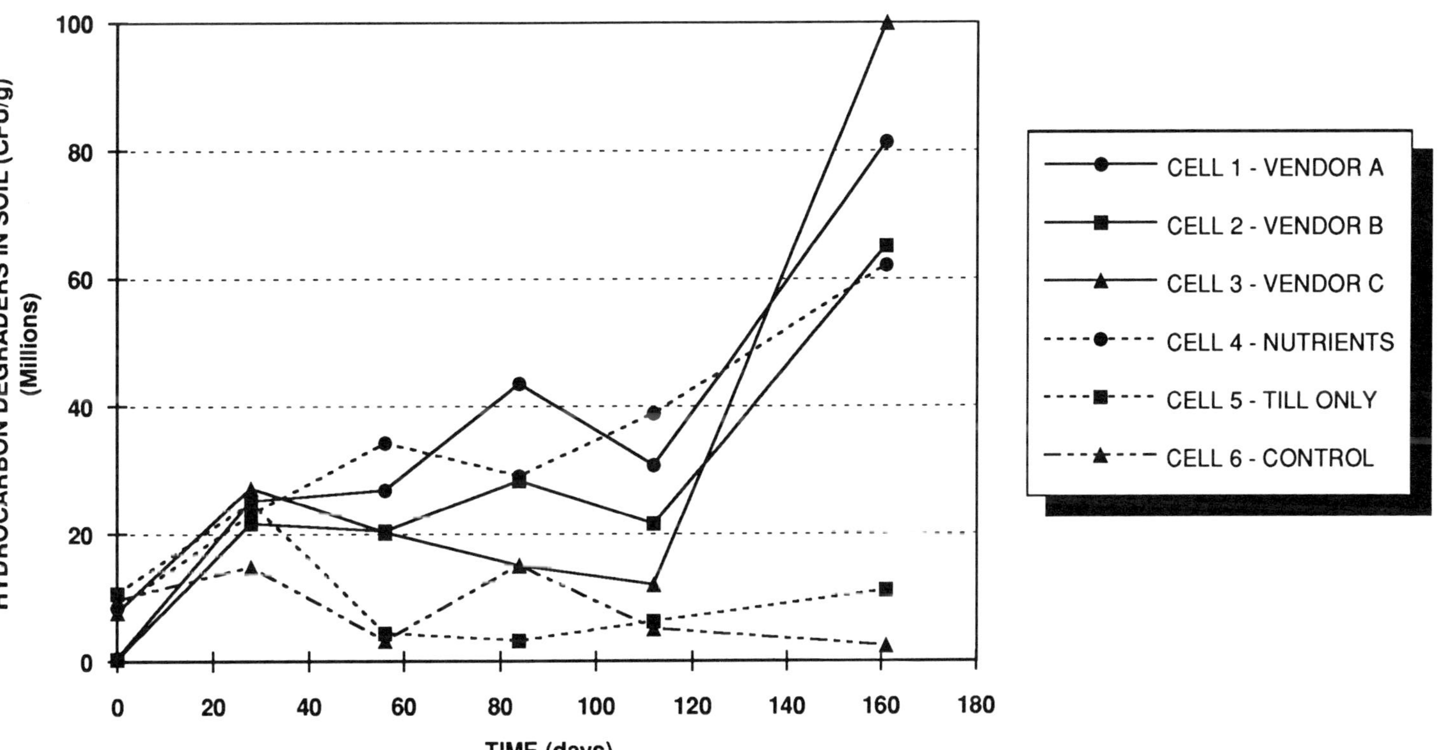

Figure 10-5. Hydrocarbon-degrading bacteria pilot test results

nitrate fertilizer to these cells. However, both the ammonia and the nitrite-nitrate concentrations decreased rapidly in Cells 2 and 4 during the first 56 days of the pilot test, suggesting that these inorganic forms of nitrogen are either quickly utilized by bacteria, or are not stable in these soils.

TKN is considered to represent the sum total of organic nitrogen and nitrogen in the form of ammonia.[11] TKN results for Cells 1 through 4 were significantly higher than the TKN results from Cells 5 and 6. This suggests that an organic nitrogen source may be used in the nutrient products from Vendor A and Vendor C. It also suggests that the nitrogen in the ammonium nitrate fertilizer that was added to Cells 2 and 4 may be quickly converted to organic nitrogen forms. It is also apparent that there are minimal concentrations (approximately 25 to 100 mg/kg) of organic nitrogen present in the soil without addition of amendments.

An increase in TKN concentrations in all of the soil cells was noted after 112 days of the pilot test. This increase was greatest in Cells 1 through 4, and may be related to a qualitative change in the nitrogen-fixing bacteria population at the onset of cold weather.

Water-Soluble Phosphate Results

The most striking feature of the water soluble phosphate results presented in Table 10-3 is the steadily increasing concentrations of phosphate measured in Cell 3 (Vendor C) from an initial concentration of 15 mg/kg to a final concentration of 500 mg/kg. It is apparent that the weekly additions of the Vendor C inoculant and nutrient solutions to Cell 3 resulted in an accumulation of water-soluble phosphate in the soil test cells. These results correspond to visual observations of a salt-like, light-colored crust on the surface of Cell 3 soils during the latter half of the pilot test.

The Vendor A nutrient product also contains a small amount of water-soluble phosphate additive. Small increases in phosphate measurements (e.g., from 8 mg/kg to 27 mg/kg between days 28 and 56) that correspond to nutrient additions are apparent in Cell 1 results. Phosphate compounds were not added to Cells 2, 4, 5, and 6. Concentrations ranging from 1 mg/kg to 12 mg/kg were measured in these soils that were not amended with phosphate.

Moisture Results

Following the initial amendment of soils, the percent moisture of test soils ranged from 5.4% to 19% throughout the pilot test (Table 10-3). Moisture in Cells 1 through 5 was generally maintained above 10% except during the warm and dry weather of October and early November. These results suggest that moisture was not limiting petroleum degradation during the pilot test. Water was added fairly regularly during September, October, and November to replace soil moisture lost to evaporation. Generally the variation in soil moisture among different cells was minimal.

pH Results

Soil pH varied from 7.1 to 9.0 throughout the pilot test (Table 10-3). Soil pH measurements in Cells 1, 3, and 4 were generally below 8.5. Soil pH from Cells 2, 5, and 6 tended to increase slightly with time. The pH measurements are somewhat higher than values often cited for optimum biodegradation (i.e., 6.0 to 8.0). However, there does not appear to be a correlation among the cells that show higher pH values with lower TPH degradation rates.

COST EVALUATION

To assess the cost effectiveness of each of the amendments, the costs of applying the various amendments to the full-scale landfarm containing petroleum contaminated soils from the UST site were estimated. Rate constants for each of Cells 1 through 5 were calculated and applied to estimate amendment specific treatment times for the full-scale landfarm soils (Figure 10-6). A target final TPH concentration of 250 mg/kg was assumed, which corresponds to the state soil cleanup guideline.[12] The cost estimates also included the price of the vendor supplied bacteria or nutrient products, the estimated treatment time, the cost associated with application of the amendments, the cost of water application and tilling of the soil on a regular basis, and oversight costs. A cost estimate was not performed for the no-tilling treatment scenario represented by Cell 6 because the final TPH concentration exceeded the initial concentration in the pilot test.

Degradation Rate Calculation

TPH degradation rates in the pilot test cell were calculated for two time periods: for the entire duration of the pilot test, and during the latter half of the pilot test. The first evaluation of TPH in soils involved calculation of first-order kinetics degradation rate constants and TPH half-lives based on the initial TPH concentration in the soils, and the final TPH concentration in each cell after 161 days. An average initial TPH concentration was calculated from the arithmetic mean of the initial TPH results from all of the test cells, to minimize the effect of high or low initial TPH concentrations on the calculated degradation rates. This approach was used to accommodate the inherent variability of TPH concentrations in soils by minimizing the apparent advantage in degradation rates in the cells with higher initial TPH concentrations.

Although the TPH data for each plot generally do not fit well within either a linear or a first-order kinetics model, it was determined to be appropriate to fit a first-order equation to the initial and final TPH concentrations. This approach is appropriate based on the recognition of first-order kinetics for petroleum biodegradation cited in the literature.[2] An assumption of first-order kinetics is also more conservative (i.e., results in longer predicted treatment times) than a linear kinetics model when extrapolating the results to concentrations of TPH that are lower than those encountered during the pilot test.

Table 10-4 presents the TPH half-lives calculated from TPH concentrations in each cell throughout the duration of the pilot test (day 0 through day 161). It is apparent from the results that the addition of commercial bacteria inoculants from Vendors A, B, and C all resulted in significantly greater degradation rates than that associated with the addition of nutrients alone as calculated for the entire duration of the pilot test. It is also apparent that the addition of nutrients produced significantly greater degradation rates than did tilling alone. It should be noted, however, that the apparent rates

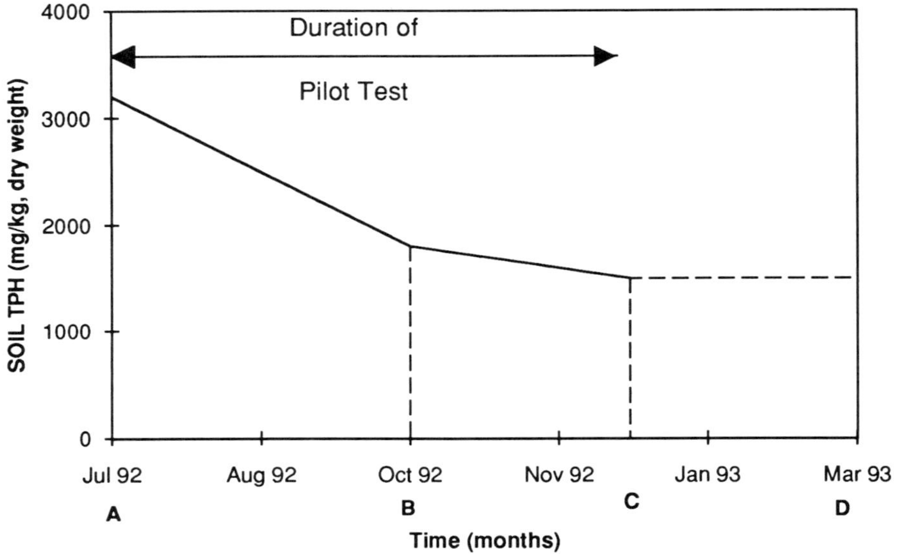

Notes:
  A to B: Removal of volatile and easily biodegraded hydrocarbons.
  B to C: Removal of more recalcitrant hydrocarbons primarily by biodegradation.
  C to D: Negligible removal of hydrocarbons due to cooler winter temperatures.

Figure 10-6.   Conceptual graph of TPH degradation.

calculated may only apply to conditions similar to those encountered throughout the pilot test. For example, as biodegradation rates are highly sensitive to temperature, the observed degradation rates are most applicable to soil temperature conditions similar to those encountered throughout the pilot test. In addition, the calculation of degradation rates from initial and final TPH concentrations includes the initial removal of easily biodegraded or volatilized compounds, and may overestimate the rate of continued hydrocarbon degradation of more recalcitrant compounds that remain.

Periodic maintenance of the full-scale landfarming was conducted throughout the performance of the pilot test, and presumably the more easily volatilized and biodegraded hydrocarbon have already been removed. Therefore, degradation rates were recalculated based on the TPH removal measured during the latter part of the pilot test from day 84 through 161 (see Table 10-4 and Figure 10-6). These degradation rates are more appropriate for application to the full-scale landfarm because landfarm maintenance has been ongoing. These results indicate that the addition of nutrients produce significantly greater apparent degradation rates than tilling alone. Interestingly, the degradation rates of the acclimated indigenous bacteria became comparable to those calculated for the vendor bacteria during the latter portion of the test period. Therefore, the benefit of the vendor bacteria was diminished as the test progressed and the indigenous bacteria became acclimated.

As full-scale maintenance and material costs are directly related to the duration of treatment, an estimate of the time required for treatment under each of the evaluated treatment options was performed. The calculation of treatment times involved applying the first-order rate constant from the TPH concentration of each plot at days 84 and 161.

An initial concentration of 3,400 mg/kg for the full-scale landfarm was assumed, based on the average concentrations measured in soil stockpile samples in April and May 1992.[13] A summary of the duration of full-scale treatment required to achieve the state soil cleanup guideline of 250 mg/kg for TPH for the five treatment scenarios is included in Table 5. Estimated treatment times ranged from 174 days for Cell 2 to 475 days for Cell 5.

Table 10-4. TPH half-lives by pilot test cell

		TPH Half-Life (Days)	
Test Cell	**Description**	**Days 0 through 161**	**Days 84 through 161**
1	Vendor A	71	63
2	Vendor B	71	46
3	Vendor C	96	73
4	Nutrients	128	69
5	Till Only	161	126
6	Control	NA	NA

Cost Estimate for Treatment

Cost estimates were prepared for application of each of the five treatment scenarios, as represented by Cells 1 through 5 of the pilot test, to the full-scale landfarm. Comparative costs for application of each of the five treatment scenarios were developed and are summarized in Table 10-5. The costs are based on future materials and maintenance required to complete treatment, and do not include the cost for landfarm construction.

Probably the most significant feature of the comparative costs is the wide variability in the cost of amendment materials. The estimated costs for full-scale use of the amendments evaluated for the pilot test ranged from \$32,000 to \$164,000. Bacteria cultures sold by vendors vary enormously in cost. The costs of applying vendor bacteria generally outweighed the financial benefit of the reduced treatment times that could result from the use of the commercial bacteria. For example, use of the Vendor A products resulted in enhanced degradation rates compared to those achievable through the use of fertilizer (nutrient) and indigenous bacteria. However, the cost of the Vendor A product far outweighs the cost benefits of its use, as demonstrated by unit costs of \$22.56 per yd^3 for treatment using the Vendor A product, in comparison to \$5.94 per yd^3 for addition of nutrients alone. Vendor B's bacteria culture appears to be the most cost effective of the vendor products. However, an additional \$13,000 in material and labor costs above the cost of nutrient addition alone would be incurred with the use of Vendor B bacteria.

The actual unit cost for treatment of the full-scale landfarm soils, based on the use of ammonium nitrate fertilizer, was calculated to be approximately \$10.50 per yd^3 of soil. This estimated actual unit cost includes approximately \$25,000 already incurred for the construction of the landfarm, subcontractor costs for maintenance, and construction management oversight.

PILOT TEST CONCLUSIONS

The results of the pilot test indicate that the application of commercially available bacteria to the contaminated soils resulted initially in greater rates of petroleum degradation than addition of nutrients alone. This conclusion is supported by the degradation rates calculated for the entire pilot test duration for the cells treated with commercial bacteria in comparison to degradation rates calculated for the cell to which only fertilizer was added. It was also demonstrated that the addition of ammonium nitrate fertilizer resulted in greater rates of petroleum degradation than the rate indicated for the till only test cell, to which no fertilizer was added. Soil aeration by tilling was also demonstrated to be necessary for enhanced petroleum degradation as shown by the comparison of the till only test cell versus the control cell results.

Removal of the petroleum from the contaminated soil by biodegradation (rather than volatilization or other removal mechanisms) was suggested by the increase in populations of heterotrophic and hydrocarbon degrading bacteria as measured in soil samples. Samples with the largest populations of both types of bacteria were associated with the test cells with the highest TPH degradation rates.

Table 10-5. Cost evaluation summary for full-scale landfarm treatment options

Pilot Test Cells	Treatment Description	Duration of Treatment (Days)	Full-Scale Treatment Costs						Unadjusted Unit Cost[a] (per yd^3)
			Water Application	Soil Tilling	Amendment Materials	Material Applications	Total Cost		
1	Vendor A	237	$8,448	$16,896	$95,237	$1,256	$121,837		$22.56
2	Vendor B	174	$6,336	$12,672	$19,874	$5,996	$44,878		$8.31
3	Vendor C	275	$9,504	$19,008	$105,300	$30,008	$163,820		$30.34
4	Nutrients	261	$9,504	$19,008	$2,324	$1,256	$32,092		$5.94
5	Till Only	475	$16,896	$33,792	$0	$0	$50,688		$9.39
6	Control	NA[b]	NA	NA	NA	NA	NA		NA

a Unadjusted for baseline cost of landfarm construction and maintenance to date, plus pilot testing. Assumes 5,400 yd^3 of soil.

b NA=Not Applicable. As the final TPH concentration exceeded the initial TPH concentration in Cell 6, a duration of treatment could not be calculated, and a cost estimate was not performed for this treatment scenario.

The use of commercially available bacteria cultures was demonstrated to produce greater TPH degradation rates than those achieved by indigenous bacteria alone, as calculated from average initial and final TPH concentrations in soil from each cell. However, at the end of the pilot test, total heterotrophic bacteria and hydrocarbon degrading bacteria population counts were elevated both in the cells with commercial bacteria cultures added and the cell with indigenous bacteria to which nutrients were added. In addition, the TPH degradation rates representing the latter portion of the pilot test (days 84 through 161) showed little significant difference between the cells with commercial bacteria added and the cell with nutrients added to indigenous bacteria. Thus, it is apparent that the advantage of adding bacteria cultures was greatest during the first few months of treatment, when the indigenous bacteria may not have been fully acclimated to the petroleum release. This advantage significantly diminished during the latter portion of pilot test treatment after indigenous bacteria had adapted to the hydrocarbon contaminants and nutrient additions. Similarly, it is likely that the period during which commercial bacteria culture additions to the full-scale landfarm would have been beneficial has passed, as periodic landfarm maintenance by tilling and moisture addition has been performed throughout the duration of the pilot test, leaving the more degradation resistant hydrocarbon constituents to be treated.

The cost of nutrient and bacteria products also has a significant impact on the cost effectiveness of the use of these products. The estimated costs for full-scale use of the amendments evaluated for the pilot test ranged from $32,000 to $164,000. The variation in costs was largely related to the material costs of the bacteria products, and secondly to the time estimated for completion of treatment.

The estimated cost of using commercial bacteria from Vendor B ($45,000) was only slightly greater than the estimated cost of amending the soil with fertilizer alone ($32,000). The time required for degradation to a 250 mg/kg TPH soil cleanup guideline was estimated to be slightly less with use of the Vendor B product (174 days) as compared with the use of fertilizer alone (261 days). However, as explained above, it is likely that the advantage of enhanced degradation rates associated with the Vendor B product may have already been diminished in the full-scale landfarm, as the indigenous bacteria in the landfarm have had nearly a year to adapt to the soil contaminant conditions.

Recommendations for Full-Scale Landfarm Maintenance

For the full-scale landfarm, the application of ammonium-nitrate fertilizer was recommended. Approximately 5,000 pounds of nitrogen in a fertilizer substrate was applied to the landfarm. The fertilizer was dissolved in approximately 20,000 gallons of water and sprayed onto the landfarm with a water truck. The soil was then mixed by tilling.

Tilling of the landfarm is being performed biweekly beginning in April 1993. Water has been added monthly, as required, immediately prior to a tilling event to maintain the soil moisture above 10%. This maintenance schedule for tilling and water addition will be maintained from April 21 through November 30, 1993, as weather permits.

It was calculated that the use of ammonium nitrate fertilizer could reduce TPH concentrations to 250 mg/kg within 261 days (Table 10-5). It was estimated that the total cost for treatment will be $32,092. Soil treatment from April 21, 1993 to November 30, 1993, would incorporate over 220 days of treatment. Therefore, the treatment of landfarm soils may be nearly complete by the end of 1993. However, as cooler temperatures can drastically affect the rate of biodegradation in soils, there will be contingency planning for continuing the treatment of petroleum contaminated soils in the spring of 1994.

REFERENCES

1.   Atlas, R.M., *Petroleum Microbiology*. MacMillan Publishing, New York, 1984.

2.   Bossert, I., and Bartha R., The fate of petroleum in soil ecosystems, in *Petroleum Microbiology*, R.M. Atlas (Ed.), MacMillan Publishing, New York, 1984.

3.   Lee, M.D., Biorestoration of aquifers contaminated with organic compounds, *CRC Critical Reviews in Environmental Control*, Vol. 18, p. 29, 1988.

4.   Miller, R.N., and Hinchee, R.E., A field scale investigation of enhanced petroleum hydrocarbon biodegradation in the vadose zone at Tyndall AFB, Florida. *Proceedings of Petroleum Hydrocarbons and Organic Chemicals in Ground Water: Prevention, Detection and Restoration*, Water Well Journal Publishing Co., Dublin, Ohio, 1990.

5.   Davis, J.B., *Petroleum Microbiology*, Elsevier Publishing, Amsterdam, The Netherlands, 1967.

6.   Litchfield, J.H., and Clark, L.C., Bacterial activity in groundwaters containing petroleum products, *American Petroleum Institute Publication 4211*, Washington, D.C., 1973.

7.   Piotrowski, M.R., Bioremediation: Experts explore various biological approaches to cleanup, *Hazmat World*, Advanstar Communications, Inc., January 1991, p. 44, 1991.

8.   Eastcott, L., Shiu, W.Y., and MacKay, D., Modeling petroleum products in soils, *Petroleum Contaminated Soils, Volume I.* Kostecki, P.T., and Calabrese, E.J. (Eds.), Lewis Publishers, Chelsea, Michigan, 1989.

9.   Senn, R.B., and Johnson, M.S., Interpretation of gas chromatography data as a tool in subsurface hydrocarbon investigations, *Proceeding of the Conference on Petroleum Hydrocarbons and Organic Chemicals in Ground Water: Prevention, Detection and Restoration*, Water Well Journal Publishing Co., Dublin Ohio, 1987.

10.   Kappeler, T., and Wuhrmann, K., Microbial degradation of the water soluble fraction of gas oil, *Water Research*, Volume 12, p. 327, 1978.

11.   American Public Health Association, *Standard Methods for the Examination of Water and Wastewater*, American Public Health Association, Washington, D.C, 1989.

12.   Colorado Department of Health, *Storage Tank Facility Owner/Operator Guidance Documents for Initial Site Characterization, Second Level Site Assessment, Use of State Cleanup Guidelines, and Management of Contaminated Materials*, December 1992.

13.   Engineering-Science, Inc., *Underground Storage Tanks Site Assessment Report and Corrective Action Plan, Defense Finance and Accounting Service - Denver Center, Lowry Air Force Base, Denver, Colorado*, May 1992.

Questions and Answers:

Q. Did the vendors provide you with some sense of the formulation of constituents, other than organisms that were in their products? For example, some freeze-dried cultures include bran, which represents an alternate carbon source.

A. Yes they did. They provided all the information on the vendor bacteria. That also includes nutrients. For instance, bran was certainly present in one particular instance.

Q. So there were some differences of the constituents, besides the organisms.

A. That's true, and there's definitely a difference in terms of the amount of labor that went into preparing a particular product for application. One was very simple, add it to water. Another was very difficult, incubate and let it bubble overnight for 24 hours. If you had to do this on a regular basis, that becomes pretty time consuming.

Q. What was in B different from A and C?

A. B certainly had the bran, as I recall, and then A was suspended in a liquid medium as opposed to freeze-dried and that sort of thing. I don't recall right off the top of my head which one differed from the other but some you applied your own ammonium nitrate fertilizer, let's say, as opposed to the other one in which the nutrients were already present in the solution.

Q. I'm Sara Tremaine from Hydro Systems; when you were doing your pilot test and you were monitoring your pots over time, did you take multiple samples in each plot?

A. Yes we did. We tried to collect a representative sample, not just one. We selected it basically on five locations, four around the perimeter and one within the center portion, and then composited that sample to try to partially eliminate the heterogeneity.

Q. So you have one data point for each time step?

A. Yes, it would be one composite sample.

Q. What I was wondering was if you used non-linear parameter estimation to get rate constants?

A. No, we used a zero order.

Q. I was wondering how the pilot test compared to the full scale at this point?

A. The first time that we were able to assess that, the rates for the full scale appeared to be somewhat slower. This may be because of the large scale nature and the inability to thoroughly mix that soil. So right now, anyway, they appear to be somewhat slower in the full scale.

Q. Are there temperature differences in terms of season?

A. Well certainly, we started the pilot test in July and went into December. We started the full scale in April and went into early December so there were some temperature differences between the pilot and full-scale.

CHAPTER 11

Bioremediation of a Diesel Fuel Spill – an Interim Report

Maryellen Martin and Dean Anson II
Anson Environmental Ltd. — Huntington, New York

INTRODUCTION

Contaminated soil and groundwater are facts of life in our industrialized society. The environmental community has been attempting to identify lower cost innovative remedial methods for dealing with this contamination. One of the techniques being investigated is bioremediation. Today, this technology is gaining acceptance and popularity because of consistent success stories and cost savings compared to other remedial alternatives.

The following is an interim report describing the currently successful use of bioremediation to cleanup soil and groundwater contaminated by an underground tank failure which released 5,000 gallons of diesel fuel.

EXECUTIVE SUMMARY

In the winter of 1989, an underground storage tank failed and 5,000 gallons of diesel fuel leaked into the soil below it. Soil borings confirmed that most of the diesel fuel was located between 17 and 65 feet below grade. Nine groundwater monitoring wells were used to monitor dissolved and floating product on site and the depth to groundwater. Groundwater is 65 feet below grade.

The information obtained from the wells demonstrated that the uppermost groundwater was perched water. This perched water was contained by clay lenses, some of which were ten feet thick. The underlying gradient of the water levels is negligible.

The topography of the site (a three to one slope), proximity of a building, roadway, parking lot and emergency generator made excavation of the contaminated soil impossible. The estimated cost of conventional pump and treat remedial techniques was too high in comparison to other alternatives available. A bench test was performed to determine the efficacy of bioremediation using indigenous bacteria and it showed that bioremediation was a viable option.

Therefore, because the fine sand, silt and clay under the site have been constraining the movement of the diesel fuel from the soil, bioremediation was the remedial alternative of choice. With the cooperation of and oversight by New York State Department of Environmental Conservation (NYSDEC) a conservative, step-wise work plan was developed.

As a preliminary to the *in situ* bioremediation, the soil was tested for total petroleum hydrocarbons (TPH), soil nutrients and colony forming units (CFU) of bacteria that are known to be petroleum consumers. Based on information from the literature, the bench test and the preliminary tests the amounts and kinds of nutrients were determined.

Bioremediation system design consisted of twelve borings with three injection points per boring. These three injection points are at levels below grade such that nutrients could be introduced in the area of soil contamination, 15 feet, 30 feet and 60 feet below grade. Each point has slotted screen at the lower two feet to allow the passage of liquid nutrients. These borings were placed throughout the area of contamination.

The first step in the work plan was to add bottled deionized water. The purpose being to keep the moisture content of the soil within the desired ranges and supply additional oxygen to the bacteria. A second round of soil testing determined that the CFU's had increased.

At the time of this writing, significant progress has been made in reducing the concentration of total petroleum hydrocarbons in the soil, quantity of diesel fuel floating on the groundwater and dissolved diesel fuel in the groundwater.

In the future, additional nutrients, including nitrogen and phosphorous, may be introduced in the same manner as the deionized water.

SITE DESCRIPTION

The site in question is located on the North Shore of Long Island, New York. The site had a 5,000 gallon underground storage tank that failed suddenly and released five thousand gallons of diesel fuel into the subsurface environment. The soil contamination at the site begins at 17 feet below the ground surface and extends to the groundwater interface which is 65 feet below grade.

Site Geology and Hydrogeology

The soils at the site are composed of fine sands, silts and clays. This soil composition does not allow for the rapid vertical movement of liquids, including rainwater. The area surrounding the spill area is impervious paved parking lots and roadways inhibiting movement associated with penetration. The unpaved area of the spill has an approximately three to one slope. Therefore, rainwater tends to runoff and not penetrate the surface soils to reach the spilled materials, approximately 17 feet below the surface.

The ground water encountered at 65 feet below grade is perched water which is contained by clay lenses typical of the subsurface conditions on the North Shore of Long Island. Comparison of surface elevations and depth to water at the wells onsite and in a Nassau County well located off site substantiate the fact that the wells onsite are not hydraulically connected to the Upper Glacial Aquifer, the uppermost aquifer under Long Island.

Nine groundwater monitoring wells were installed surrounding the spill location. Clay layers approximately ten feet thick, were encountered during the drilling process of some of the monitoring wells. The sporadic clay lenses lent evidence to the hypothesis that the water in the wells is perched water retained by clay lenses. The underlying gradient of the water levels in the monitoring wells is negligible. For instance, the difference in the depth of water between the wells at opposing ends of the spill site is .036 inches, and the wells are 200 feet apart.

Soil and Groundwater Sampling

The groundwater monitoring wells onsite are sampled quarterly and submitted for laboratory analysis via EPA method 601 and 602. The constituents of the spill, benzene, toluene, ethylbenzene and total xylenes (BTEX), have been detected only in the two wells (MW5 and MW8) in the immediate vicinity of the spill.

Soil samples are collected from 15-17 feet, 30-32 feet and 55-57 feet below grade and are analyzed for total petroleum hydrocarbons and CFUs periodically. All sampling is performed using US Environmental Protection Agency and NYSDEC sampling protocols for quality control and quality assurance.

Summary

Based on the geology, hydrogeology and results of the groundwater samples, it was determined that the vertical and horizontal migration of the groundwater plume is limited to the immediate area of the former tank location. The majority of the contamination is bound in the fine sand, silt and clay indigenous to the site. Before bioremediation could be recommended as a remedial technique, a bench test had to be performed to determine if the indigenous bacteria and other microbes onsite would degrade the diesel fuel released.

BENCH TEST

The first step in determining the viability of bioremediation is to conduct a bench test. Contaminated soil and water were collected from the site and were subjected to an *ex situ* culturing test using the following protocol.

Nine polypropylene test bins were established to conduct this bench test. Three groups were established - the first as a control (no treatment), the second as a nutrient treatment only and the third as nutrient treatment with the addition of bacteria from clean Long Island soils. All groups were placed in a refrigeration unit to maintain the temperature at 50-60 degrees Fahrenheit in order to mimic *in situ* conditions. Each group was replicated three times for testing purposes.

The containers were air tight, but the nutrient mix was added to Groups B and C on a weekly basis. The control group was also opened for a time equal to the others each week (approximately one minute). Dissolved oxygen was administered monthly. Both the water fraction and soil fraction were tested monthly from each group for total petroleum hydrocarbons and volatile organic compounds, including benzene, toluene, ethylbenzene and total xylenes. The soil and water samples were tested for three months.

The bench test clearly demonstrated that treatment with nutrient mix would likely cause a substantial decrease of total petroleum hydrocarbon levels at the subject site. The test results demonstrated the greatest reduction in Group B. Total petroleum hydrocarbons (TPH) were reduced from 880 parts per million (ppm) in the soil to 10 ppm. Water TPH levels were reduced from 740 ppm to less than 1.5 ppm (the detection limit of the laboratory). Further, the containers were kept steady, had limited head space and were opened only once per month, thus limiting volatilization losses.

In summary, the best results were achieved with the addition of nutrients only to the original soil samples in Group B. The control group (Group A) and Group C achieved lesser reductions in TPH but less significantly than the samples to which the nutrients were added. This led to the decision to add nutrients and additional sources of oxygen to the contaminated soil onsite.

BIOREMEDIATION SYSTEM DESIGN

Based on the successful outcome of the bench test a bioremediation system was designed for the site. Twelve borings with three injection points in each boring were placed using a four foot square grid in the former tank location. The injection points extended 15 feet, 30 feet and 60 feet into the ground. The points were made of 1.5-inch diameter PVC piping with the lower 2 feet made of slotted screen to allow liquids to enter the soil medium.

Starting in April 1992, one-third of a gallon of deionized water was added to each point, three times per week. During the time that the water was added the NYSDEC reviewed recommendations for nutrients to be added to the soil.

The NYSDEC bioremediation protocol for sampling soil and groundwater included the following:

- Collect soil samples prior to the addition of nutrients. Analyze
  samples for TPH, soil nutrients and colony forming units (CFU).

- Collect periodic soil samples and analyze for TPH and CFU.

- Collect weekly depth to groundwater and floating product data.

- Collect and analyze quarterly groundwater samples for volatile
  organic compounds.

FINDINGS

The addition of deionized water to the soil stimulated the growth of bacteria and microbes in the soil as is indicated in Figure 11-1. The number of CFU of total bacteria have increased at all three depths sampled (15 feet, 30 feet and 55 feet below grade). In addition, the number of <u>Pseudomonas aeruginosa</u> has increased as well as the number of <u>P. putida</u>, which has recently become more active onsite.

Figure 11-2 illustrates that at the three depths sampled, the concentrations of TPHs in the soil has decreased during the increase in the total bacteria populations. It is deduced that the bacteria are using the petoleum hydrocarbons as a source of nourishment and breaking the more complex chains into simpler carbon units.

The bacterial action has also reduced the quantities of diesel fuel floating on the groundwater (Figure 11-3) and dissolved benzene, toluene, ethylbenzene and total xylenes (BTEX) in the groundwater (Figure 11-4).

CONCLUSIONS

To date, bioremediation has been successful in cleaning up the contamination caused by the discharge of diesel fuel at the Lake Success, Long Island, New York site. Increases in the amount of moisture in the soils have stimulated bacterial and microbial activity. Bacterial and microbial action has reduced the concentrations of total petroleum hydrocarbons in the soil from depths of 15 to 65 feet below grade. The microbes and bacteria also reduced the concentrations of dissolved benzene, toluene, ethylbenzene and total xylenes in the groundwater. As a result, the quantity of floating product has also been reduced.

Although bioremediation as a remedial alternative is in its infancy, figures published in March 1991 by the USEPA indicate that bioremediation projects are being considered, planned, or implemented in more than 140 sites, including projects at 22 sites listed on the Federal Superfund's National Priorities List (NPL).

As the number of success stories continue to increase, responsible parties and regulatory agencies will become more accepting of bioremediation as a remedial alternative. At least, New York State is continuing to develop standard bioremediation protocols on a site specific basis.

  Principles and Practices for Diesel Contaminated Soils, Vol III

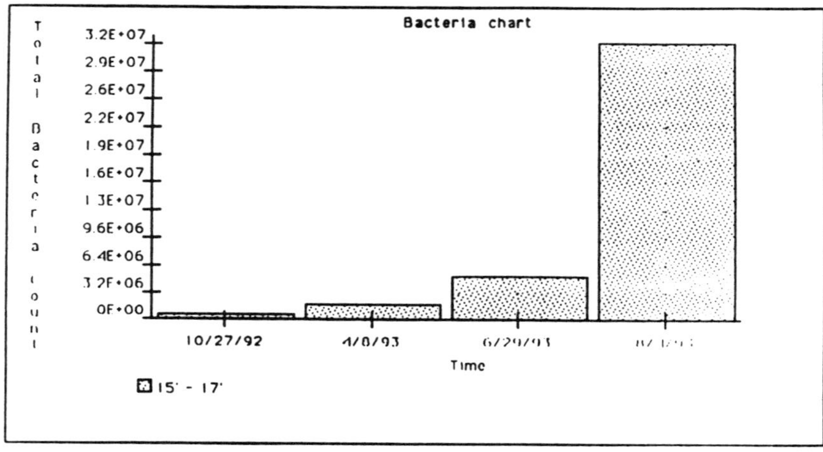

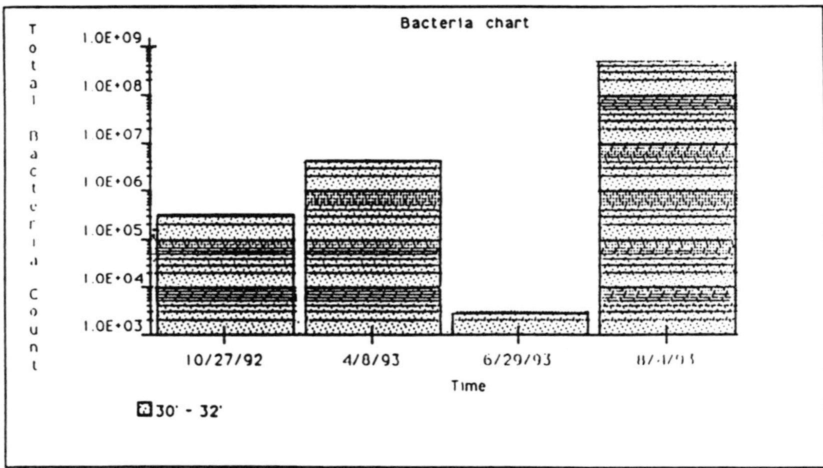

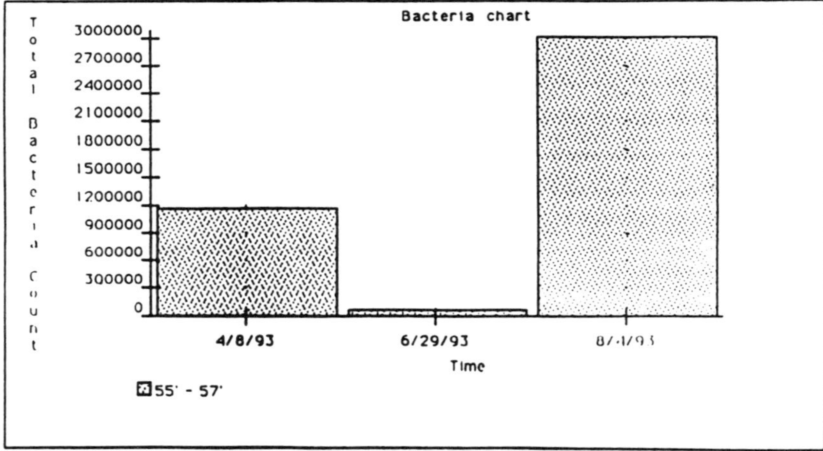

Figure 11-1 Total Bacteria Counts in Colony Forming Units (CFU)

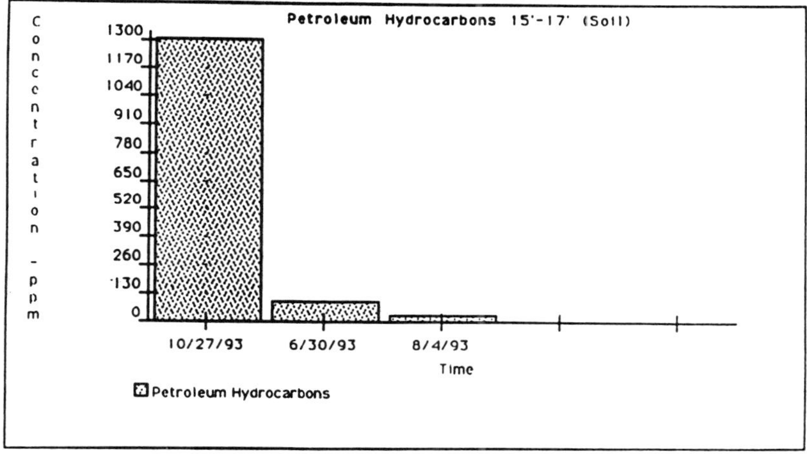

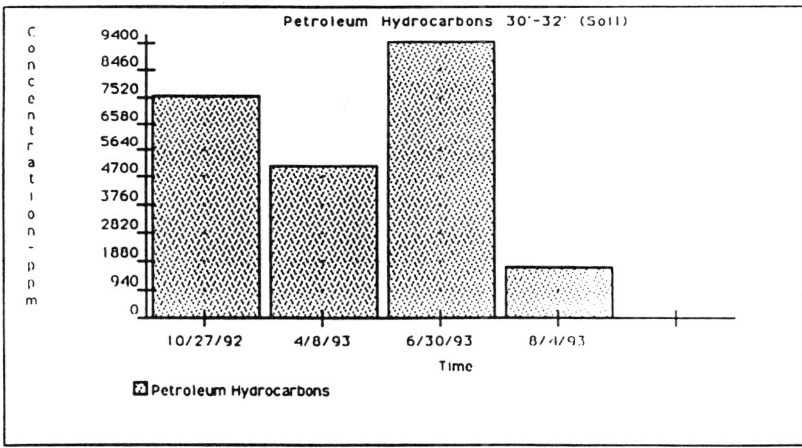

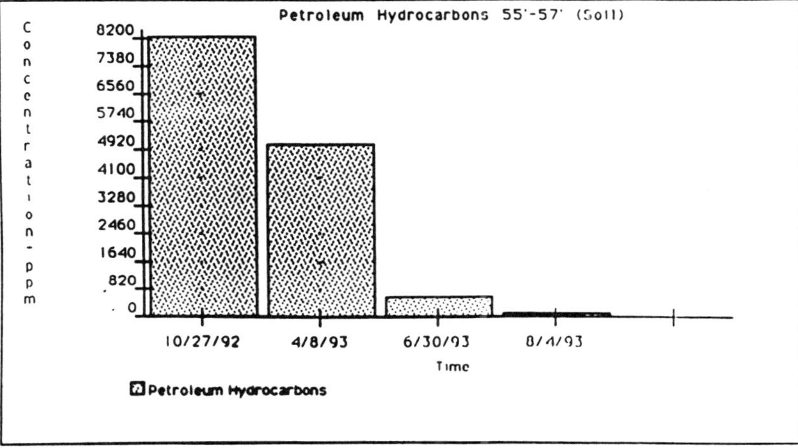

Figure 11-2 Total Petroleum Hydrocarbons (TPH) in Soil

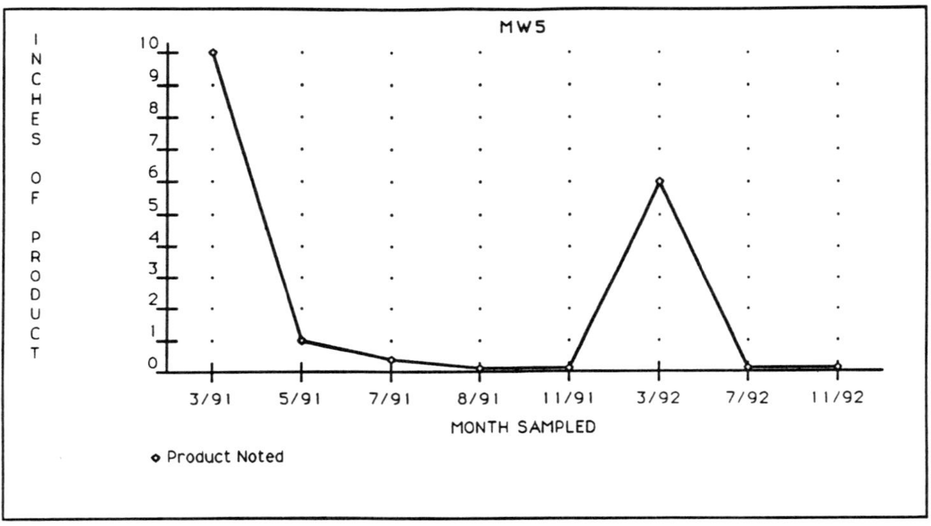

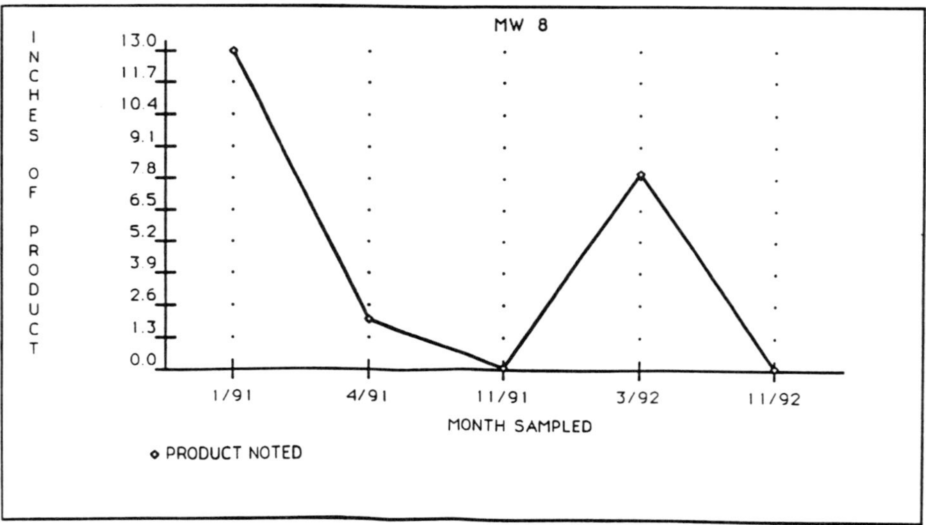

Figure 11-3   Quantities of Floating Diesel Fuel on Groundwater

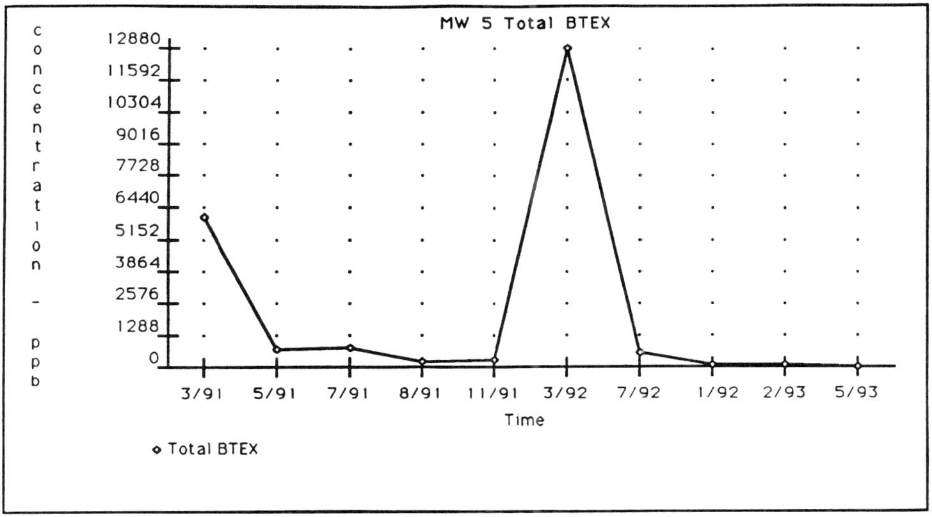

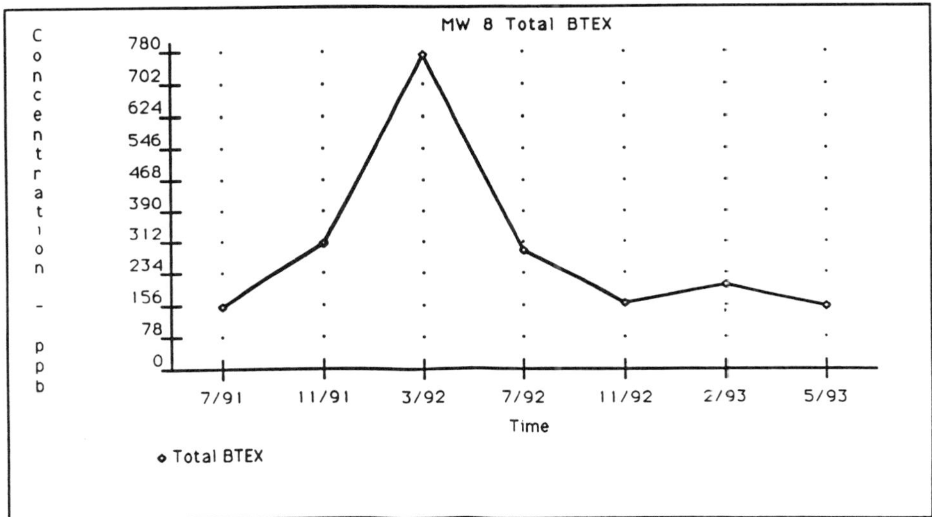

Figure 11-4 Total Benzene, Toluene, Ethylbenzene and Xylene (BTEX) in Groundwater

CHAPTER 12

An Investigation of Potential Methods to Enhance *In Situ* Bioremediation of Diesel Fuel

Greg D. Naugle and Robert J. Sterrett, Ph.D.
Hydrologic Consultants, Inc.

John W. Meldrum and Melvin L. Burda
Burlington Northern Railroad

INTRODUCTION

This report summarizes a two year investigation of methods of enhancing the *in situ* bioremediation of diesel range hydrocarbons. The work was conducted at a railroad maintenance and refueling yard that is located in Denver, Colorado (Figure 12-1) by Hydrologic Consultants, Inc. (HCI), in conjunction with the Colorado School of Mines (CSM). Ultimately, the intent of the study was to identify a cost-effective method of remediating diesel hydrocarbons that have spilled and migrated into the subsurface. Diesel fuel spills are common in rail yards during engine repair, fueling activities, as well as remote train derailments sites. In these situations, a method of remediation is needed that requires a minimal amount of disturbance to ongoing railroad operations. Currently, most bioremediation methods involve the excavation of soil, but these methods are impractical in active or ecologically sensitive sites due to the numerous logistical, engineering, and economic problems associated with excavating soil. Consequently, an *in situ* method of bioremediation that would reduce the concentrations of diesel hydrocarbons in soil would be beneficial for these types of situations.

The use of enhanced bioremediation for the treatment of hydrocarbons in the vadose zone is a relatively new technique[1]. The majority of previous studies that tested enhanced bioremediation treatment techniques focused on land farming, a method that requires the physical excavation of the soils containing hydrocarbons and a large treatment area to allow microbial populations to oxidize the hydrocarbons. The present study of enhanced bioremediation, on the other hand, focused on a combination of soil vapor extraction with nutrient or surfactant injection into the subsurface to stimulate the naturally occurring biodegradation of diesel hydrocarbons. This method of enhanced bioremediation, therefore, involves a minimal removal of soil and a minimum amount of space to be implemented.

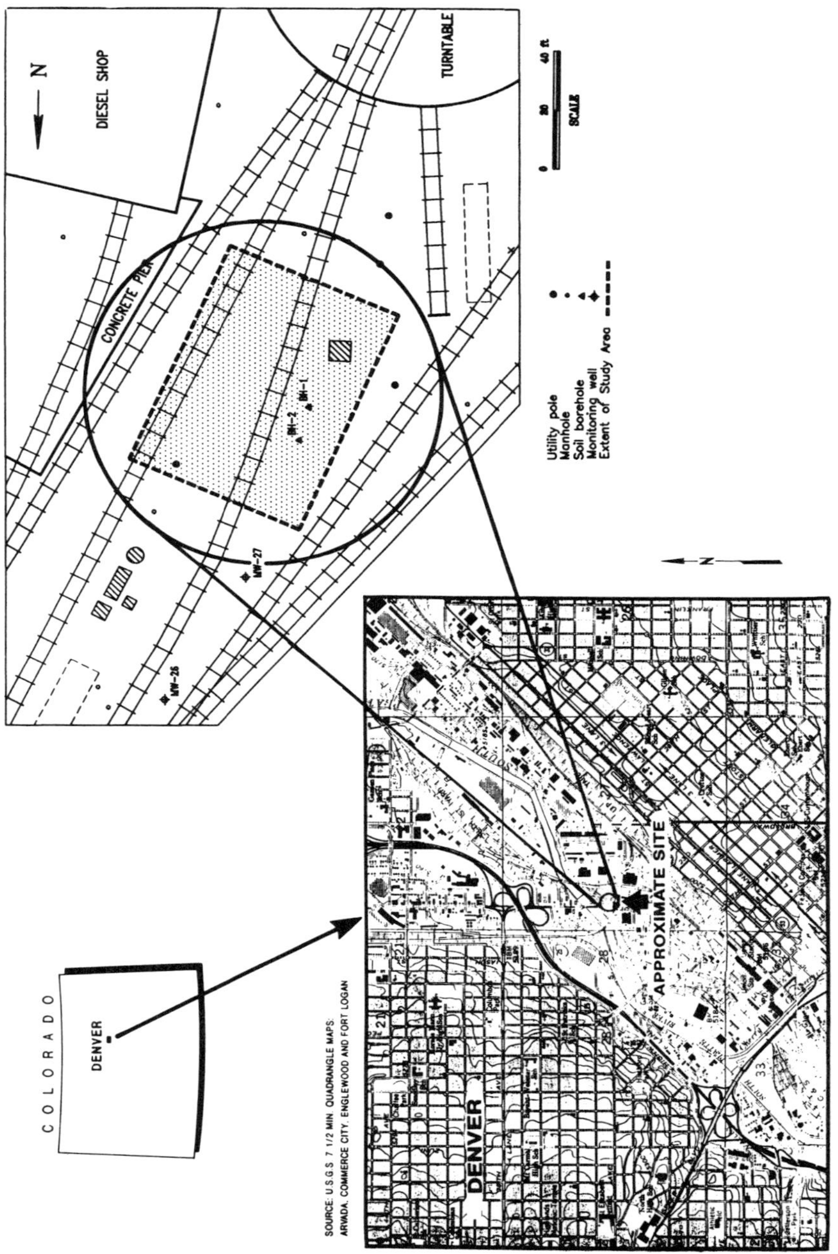

Figure 12-1 Site Location Map

Background

A vapor extraction system (VES) is based on the concept that volatile organic compounds (VOCs), or chemicals with high volatilization potentials, within soils will evaporate at ambient temperatures if air flow in the subsurface is induced. The use of VES has been shown to be very effective at remediating soils containing VOCs at a number of sites[2]. However, a VES is not as efficient for remediating heavier molecular weight petroleum hydrocarbons, often found in diesel fuel, because these compounds do not exhibit high volatilization potentials (vapor pressures). For these heavier molecular compounds, oxidation by naturally occurring microbes in the subsurface can be an important degradation mechanism if the site specific soil conditions facilitate the process. In most cases, the growth of naturally occurring microbes will be stimulated by the increased flow of oxygen that is produced by a VES, and thus VES can be a viable remediation technique for soils containing diesel ranged hydrocarbons.

There are several factors that can influence the success of using a VES for the remediation of diesel hydrocarbons. For example, low moisture contents can inhibit the growth of the *in situ* microbes necessary for the bioremediation process. Site specific-conditions, including naturally low soil moisture contents often limit the effectiveness of bioremediation processes[3]. Additionally, long term operation of a VES reduces the moisture content of the site soils, which further reduces the effectiveness of naturally occurring bioremediation processes. In these cases the growth rate of the *in situ* microbes can be enhanced by increasing the soil moisture contents at the site.

The lack of a proper nutrient balance in the site soils is another condition that can inhibit microbial growth at a particular site. The soil found at a particular site may be deficient in a number of nutrients necessary to stimulate microbe growth, including: nitrogen, phosphorous, potassium, sulfur and other trace elements[4]. The reintroduction of these nutrients into the soil has been shown to increase the biodegradation rate of hydrocarbon compounds. For example, land farming of a jet fuel contaminated soil, enhanced by supplementing the soils with nitrogen, produced a greater than 7,000 fold decrease in the observed concentration of benzene[5]. Laboratory experiments assessing the enhanced biodegradation of polycyclic aromatic hydrocarbons (PAH) indicated that the addition of nutrients increased the overall biodegradation rates between 12 and 68 percent depending on the specific PAH compound[3].

Some investigators[6-8] have advocated the use of surfactants to enhance the biodegradation of hydrocarbons. Surfactants possess several desirable characteristics that can enhance the bioremediation process, including:

- the ability to alter the wettability of the hydrocarbon compounds,

- the ability to reduce the surface tension of the hydrocarbon compounds, and

- the ability to produce a hydrocarbon emulsion.

Together, these characteristics tend to decrease the adsorption of hydrocarbons to soil particles and increase the availability of the hydrocarbons to the microbes. Both of these processes accelerate metabolism of the hydrocarbon compounds by the microbes within the subsurface.

Goals Of The Present Investigation

The primary objective of the current investigation was to utilize the most recent innovations in bioremediation technology (described in the previous section) in an active rail yard environment. A criterion for the study was that the operation and maintenance of the systems would not hinder ongoing railroad activities. The three methods of bioremediation that were tested included:

- Vapor extraction,

- Vapor extraction with nutrient injection, and

- Vapor extraction with surfactant injection.

An additional goal of the investigation was to design the system with readily available materials so that if a full scale operation was undertaken, the operation would not require specialized equipment.

Scope Of The Investigation

The investigation was conducted over a two-year period. During the first year, numerous laboratory experiments and design work were conducted. Pilot scale field operations were conducted from August through November 1991, when operations were suspended due to freezing conditions. All three methods of enhanced bioremediation, in three separate test plots, were utilized during the first year of pilot scale testing. A control test plot was also established to assess the potential biodegradation of hydrocarbons in the absence of any active remediation. Two complete sets of soil samples were collected from each of the four test plots during 1991. Due to the relatively short period of pilot scale operations achieved during 1991, an additional year of operation of the system was undertaken.

The second year of pilot scale operations began in May 1992 and were concluded in November 1992. Six sets of soil samples were collected during the second year of operations to provide a set of data that could be analyzed, using statistical methods, to assess the decrease in the total petroleum hydrocarbon concentrations resulting from the various enhanced biodegradation processes. During the second year of operations, it was necessary to abandon both the control and the original surfactant test plots due to the installation of a pipeline at the railroad yard. Surfactant injection was considered a more promising form of enhanced bioremediation, therefore, it was decided to convert the vapor extraction only test plot that existed during the 1991 operations into a surfactant injection test plot. As a result, the test plots in 1992 consisted of nutrient injection and surfactant injection test plots.

SITE LOCATION AND GEOLOGIC CHARACTERIZATION

The site selected for the investigation is located within the former diesel fueling area at the a railroad maintenance yard (Figure 12-1). Soils at the former diesel fueling area were known to contain significant quantities of diesel fuel, and therefore, this location was considered ideal for the bioremediation study. In order to minimize the possible impacts of the investigation on ongoing railyard activities, the study area was limited in size (60 ft x 80 ft). Initially, four (20 ft x 20 ft) test plots were established within this study area as shown on Figures 12-2 and 12-3.

Initial site characterization of the study area was conducted in May 1991. This initial characterization involved obtaining soil samples from approximately 63 borings. The soil borings were excavated using a 2 inch split spoon sampler and a slide hammer, and the borings ranged in depths between 8 and 20 feet. The borings were located throughout the study area with the majority being located or. a regular grid within the test plots. Soil samples were collected at various depths in the borings to obtain preliminary data for assessment and design, including:

- the geology of the study area,

- the *in situ* nutrient levels and initial moisture contents of the site soils,

- the *in situ* microbial populations in the site soils, and

- the initial total petroleum hydrocarbon (TPH) concentrations.

General Geology
The railroad yard is located along the Holocene to Pleistocene aged South Platte River terrace[9]. As such, the majority of the unconsolidated geologic deposits beneath the site are alluvial or fluvial in nature. The remainder of the unconsolidated deposits consist of miscellaneous fill associated with urban development of the area. The thickness of these unconsolidated materials vary greatly and range from zero to greater than 70 feet in thickness throughout the Denver metropolitan area[10].

Stratigraphically underlying the unconsolidated geologic deposits is the Denver formation, which is late Cretaceous to early Tertiary in age. The Denver formation is considered to represent the shallowest bedrock unit in the Denver metropolitan area[11]. The Denver formation consists of interbedded claystones and siltstones, with occasional sandstone stringers of limited areal extent. In the vicinity of the railroad yard, the dip of the Denver formation is relatively shallow and to the northeast.

Geology Of The Study Area
Based on the geologic description of the soil borings excavated during the initial characterization of the study area, the subsurface geology of the study area was subdivided into six lithologic units. These six units, in order from shallowest to deepest, are:

1)   railroad ballast,

2)   fill,

3)   high plasticity clay,

4)   well sorted fine grained sand,

5)   upper gravelly sand, and

6)   lower gravelly sand.

As the study area was located within an active rail yard, ballast was found from the surface to approximately one foot below land surface. Fill was present beneath the railroad ballast ranging in thickness from 1 to 6.5 feet. The fill consisted of sand, wood

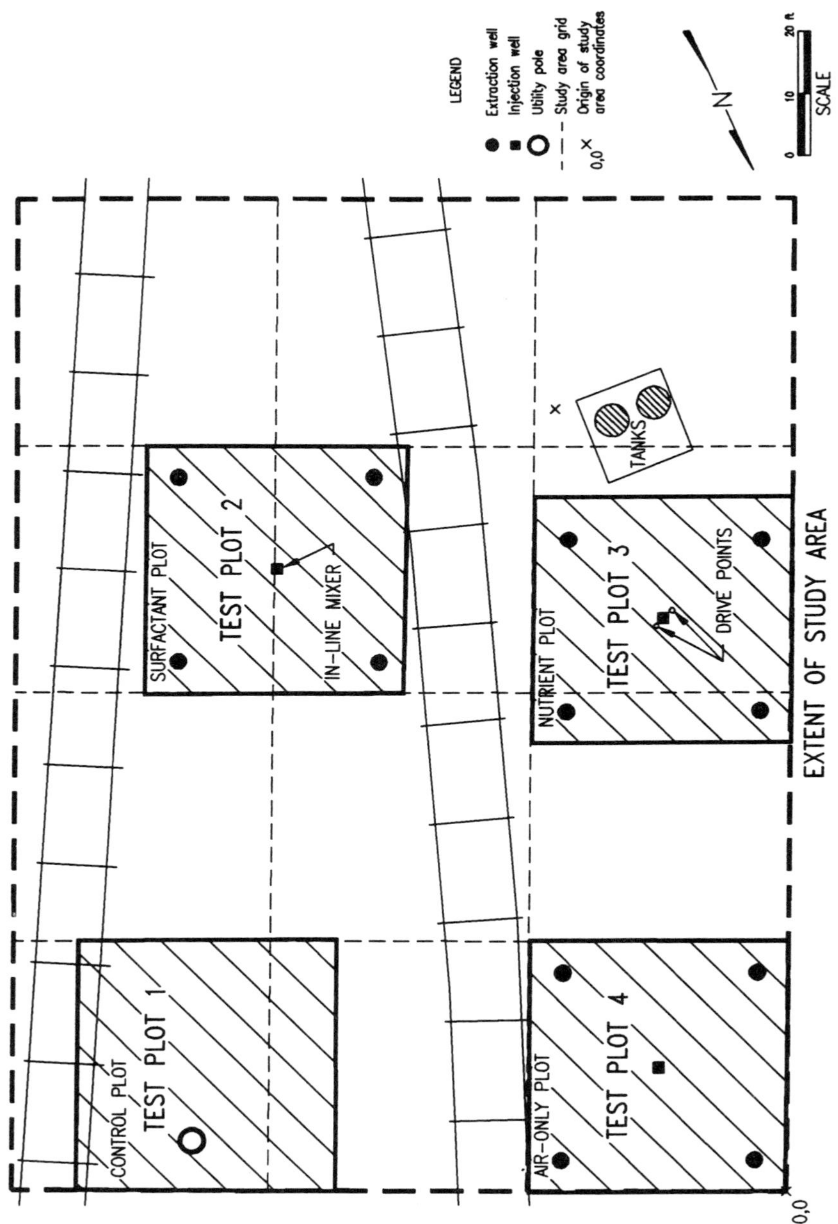

Figure 12-2   1991 Test Plot Locations

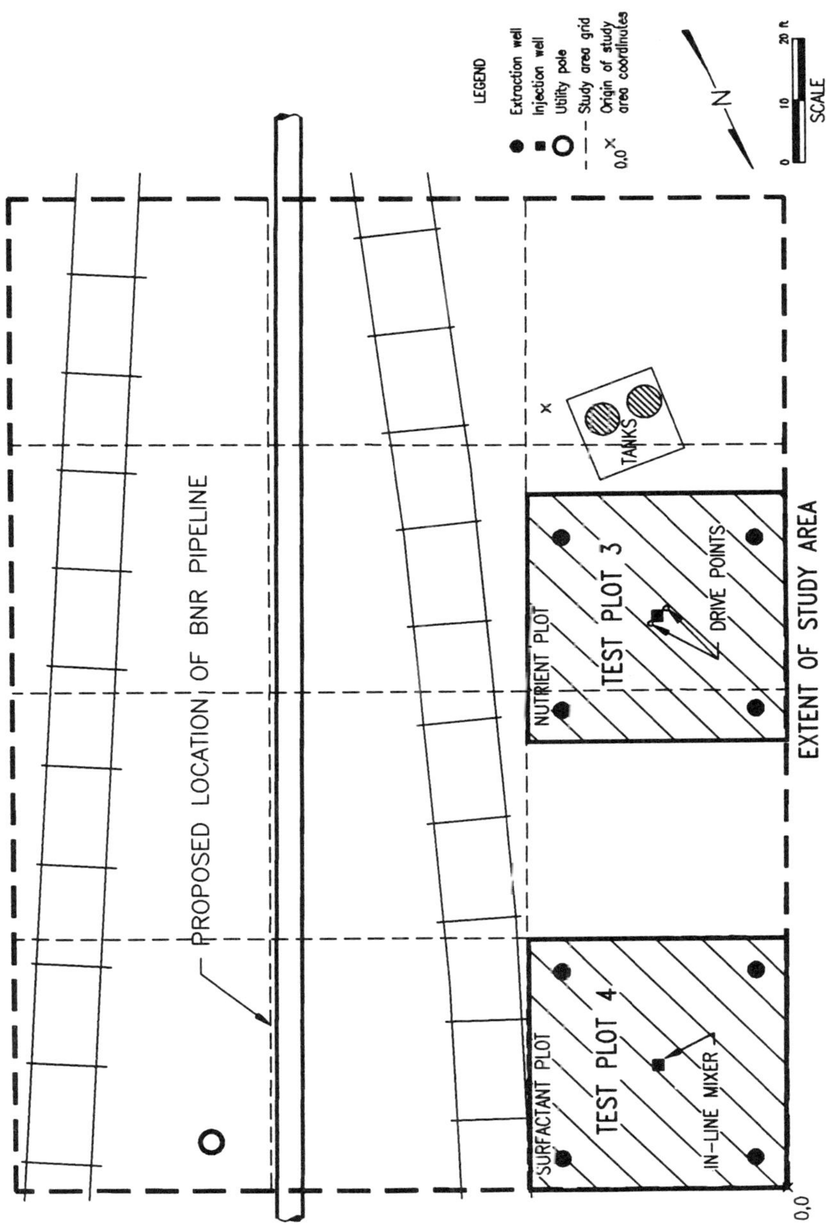

Figure 12-3   1992 Test Plot Locations

fragments, bricks, and coal fragments. Hydrocarbons were commonly observed within the fill which exhibited a strong hydrocarbon odor.

Beneath the fill, a high plasticity clay is present over the majority of the study area at depths of 4 to 8 feet below the land surface, depending on the thickness of the fill. This high plasticity clay is identified by a consistent gray color, as well as a significant silt content. The clay appears to be thickest at the northeast portion of the study area and is absent along the south and west boundaries of the study area. In most boreholes, diesel and a more viscose hydrocarbon substance were observed at the contact between the clay and the overlying fill.

Within the high plasticity clay, a thin, well sorted fine grained sand is occasionally observed. This sand ranges from 0.25 to approximately 2 feet in thickness. The fine grained sand layer consists of moderately to well sorted clasts that are predominately quartz in mineralogy, with minor amounts of other minerals. In almost all cases, samples from this sand unit exhibited a strong hydrocarbon odor. The absence of this sand layer in a number of boreholes indicates that it may represent a paleo-stream channel.

The upper and lower gravelly sand units have very similar physical characteristics, and are found at depths of approximately eight to ten feet below land surface. The difference between the two gravelly sand units is based on the clast size in the gravel fraction. The upper gravelly sand generally has smaller sized clasts, of which less than 15 percent are of the pebble (2-4 mm) and granule (4-64 mm) size; whereas, the lower gravelly sand has larger clasts. In the lower gravelly sand, approximately 20 percent or more of the clasts are in the pebble sized fraction (2-4 mm) and in the granular sized fraction (4-64 mm). Diesel fuel and more viscous hydrocarbons were generally observed in both gravel zones to depths of approximately 20 feet. The water table is encountered at a depth of approximately 24 feet below land surface, within the lower gravelly sand.

Based on the preliminary interpretation of the geology of the study area, as well as the initial TPH concentrations, the upper and lower gravelly sand units were selected for treatment. Preliminary estimates of air permeabilities indicated that a VES in the gravelly sand units would produce sufficient air flow in the vadose zone to enhance biodegradation of diesel fuel. Additionally, the gravelly sand units are confined by the low plasticity clay over most of the study area, which provides a natural cap that would increase the lateral extent of the influence of the VES.

LABORATORY EXPERIMENTS

The soil samples obtained during the initial characterization of the study area were submitted to the Colorado School of Mines (CSM) for laboratory analyses. The soil samples were then tested in the laboratory in order to measure:

• initial microbe populations, and

• initial nutrient and moisture contents of the soils.

In addition to the above experiments, a series of laboratory experiments were conducted that tested several commercially available surfactants for:

- compatibility with the site specific microbe populations,

- ability to emulsify diesel fuel, and

- performance with regard to foam generation.

The results of the surfactant experiments were then used in combination with the manufacturer's published chemical characteristics to select an appropriate surfactant for the investigation.

Laboratory Measurement Of Microbe Populations

A measure of the initial microbe populations was needed to assess whether viable microbial populations were present in the site soils. Post treatment measurements of the microbe populations would later be compared to these initial measurements in order to assess whether microbial population increases were produced by the various bioremediation methods. Microbial populations were measured in the laboratory using the viable plate count method. This method relates the microbial population to the number of colony forming units (CFU) observed.

The viable plate count technique involved mixing one gram of soil with approximately nine milliliters (ml) of a 0.1 percent sodium pyrophosphate solution in order to separate the microbes from the soil. The resulting solution was then diluted and placed in a Peptone-Tryptone-Yeast-Glucose (PTYG) agar in order to simulate the oxygen deficient environment normally found in the subsurface. After a three week incubation period, the number of CFU were counted to estimate the microbial population. Inherent to this method of estimating the microbial populations is the assumption that the CFU measured are directly correlated to the number of hydrocarbon degrading microbes that are present in the soil. This appears to be a valid assumption based on work by Hickman and Novack[12].

The results of CFU counts for the soil samples from each of the test plots are presented in Table 12-1. Post treatment measurements are also presented in Table 12-1 for comparison purposes and are discussed later. Throughout the study area (all four test plots), very little difference between the initial microbe populations was observed spatially, indicating that the entire study area had approximately the same microbial populations. However, the observed microbial populations did vary with depth.

Microbial populations were observed to decrease with depth in the site soils, probably as a result of the decreasing availability of oxygen with depth. Measured microbial populations gradually decreased and reached a minimum at a depth between 10 and 14 feet below the land surface. There was a slight increase in microbial populations in the soil samples collected from depths of 17 to 20 feet, indicating that microbial production might be influenced by the groundwater table (located at a depth of approximately 24 feet below land surface). Seasonal fluctuations in the water table, or the presence of a capillary fringe at the water table, may locally promote microbe growth in the site soils near the water table.

None of the observed microbial populations were indicative of an active soil, in that

Table 12-1 Pre and Post Treatment Microbial Populations (CFU Count Data)

Test Plot	Depth(ft)	Pre-Treatment Samples (CFU/g)			Post-Treatment Samples (CFU/g)		
		Mean[1]	Std Err[2]	No. Obs[3]	Mean[1]	Std Err[2]	No. Obs[3]
1 Control	0 to 3	3.55×10^6	6.92×10^0	4	NR	NR	NR
	7 to 8	9.12×10^4	2.82×10^0	5	5.37×10^4	4.90×10^0	4
	10 to 14	7.59×10^3	2.95×10^0	7	7.94×10^3	1.86×10^0	4
	17 to 20	9.77×10^5	3.16×10^0	4	NR	NR	NR
2 Surfactant	0 to 3	1.07×10^5	1.82×10^0	4	NR	NR	NR
	7 to 8	2.69×10^5	1.86×10^0	6	8.13×10^6	1.38×10^0	4
	10 to 14	1.12×10^4	1.78×10^0	5	2.04×10^4	2.09×10^0	4
	17 to 20	1.82×10^4	5.13×10^0	4	NR	NR	NR
3 Nutrient	0 to 3	4.17×10^5	2.88×10^0	4	NR	NR	NR
	7 to 8	1.20×10^5	8.71×10^0	5	4.68×10^6	1.91×10^0	4
	10 to 14	1.00×10^4	5.37×10^0	6	4.37×10^7	1.55×10^0	4
	17 to 20	6.76×10^4	7.24×10^0	4	NR	NR	NR
4 Air Only	0 to 3	2.40×10^6	1.95×10^0	4	NR	NR	NR
	7 to 8	1.48×10^5	1.05×10^1	5	2.82×10^6	3.24×10^0	4
	10 to 14	7.59×10^3	2.45×10^0	7	6.61×10^6	2.24×10^0	4
	17 to 20	1.45×10^4	3.72×10^0	4	NR	NR	NR

Notes: All microbial population data were obtained during the 1991 field season.

All results are reported as log-transformed values.

The nominal treatment zone was located between 8 and 15 feet below land surface.

NR indicates that no sample was obtained during the post treatment sampling, because these sample depths were considered to be outside the zone of treatment.

[1] Arithmetic average of the CFU data.   [2] Sample standard deviation of the CFU data.   [3] Number of laboratory measurements.

CFU=Colony Forming Units

observed microbial populations did not exceed background measurements reported in the literature. Hickman and Novack[12] examined the bacterial growth in three different soils types and their initial bacterial counts (8 x 10^5 CFU/g) were on average eight times greater than those found in the study area (1 x 10^5 CFU/g). Based on the results of their experiment, Hickman and Novack concluded that the majority of the microbe populations in the vadose zone were dormant and may be activated by increases in soil moisture and oxygen. Despite the relatively low initial microbe populations observed for the study area, it was concluded that the study area would benefit from the application of enhanced bioremediation techniques.

Nutrient And Moisture Content Determination

In order to assess which nutrients would stimulate the growth of the site specific microbes, a series of respirometer experiments were conducted with soil samples obtained during the initial characterization of the study area. A respirometer experiment involves the mixing of 50 grams (scg) of soil with 180 milliliters (ml) of air in a sealed serum bottle, and then allowing the bottle to incubate for a specified period of time. The growth of microbe populations consumes a portion of the initial oxygen sealed within the bottle. The observed decrease in oxygen content is then directly related to the growth of the microbes.

Based on previous work conducted by Raymond et al.[4], six combinations of nutrients were tested by adding the nutrients to the soil samples used in the respirometer experiments. These nutrient combinations included:

- nitrogen;

- nitrogen and phosphorous;

- nitrogen, phosphorous, and magnesium;

- nitrogen, phosphorous, magnesium, and calcium;

- nitrogen, phosphorous, magnesium, calcium, and iron.

For each of the nutrient combinations listed above, as well as a control group that lacked any additional nutrients, a set of 14 respirometer samples were prepared. At various times throughout a 500 hour testing period, one respirometer sample for each nutrient combination and one from the control group were tested for oxygen consumption. Upon completion of the experiment, 14 measurements of oxygen consumption were completed (Figure 12-4). An increase in oxygen consumption by a microbe population during the experiment was interpreted to indicate which nutrients were most beneficial for promoting the growth of the microbe populations in the site soils.

The results of the respirometer experiments indicated that the addition of nitrogen increased oxygen consumption by approximately ten percent. The addition of nitrogen and phosphorous increased oxygen consumption by an additional ten percent, but that the addition of other nutrients did not influence the oxygen consumption by an appreciable amount (Figure 12-4). Based on the results of the respirometer tests, diammonium phosphate was selected as the commercially available fertilizer that would be used as a nutrient source for the soils in this study.

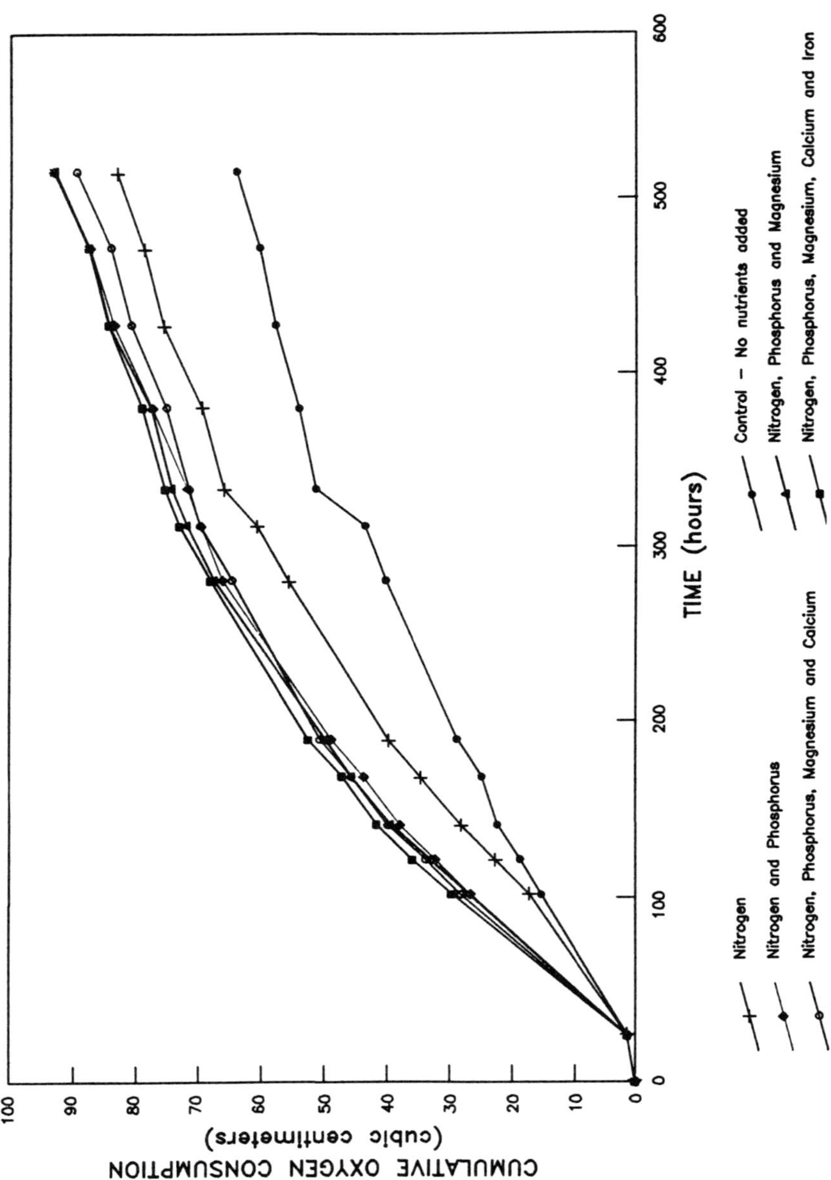

Figure 12-4   Benefits of Nutrient Addition to Microbe Populations

Two additional results were obtained as a result of the respirometer experiments: the peak daily oxygen consumption and the optimal soil moisture content for microbe growth. Based on the initial site soil moisture and the respirometer experiments, it was found that the optimal soil moisture content was approximately ten percent. This soil moisture was then used with the results of the nutrient experiments to calculate that a 0.5 gram per milliliter (g/ml) mixture of diammonium phosphate fertilizer and water, when injected to the vadose zone, would provide the optimal nutrient and moisture contents for microbe growth in the soils. The respirometer experiments also indicated that there was an upper limit of oxygen consumption by the microbial populations, and the corresponding air flow rate that would meet this oxygen demand was calculated to be approximately six cubic feet per minute (cfm) for the size of the test plots.

Laboratory Screening Of Commercial Surfactants

Two sets of screening criteria were used to select an appropriate commercially available surfactant for the investigation. The first set of criteria were based on the manufacturer's published chemical characteristics, and the second set of screening criteria were based on the surfactant's performance in a series of experiments.

The manufacturer's chemical characteristics were used to reduce the number of surfactants that would be tested in the laboratory. For example, cationic and anionic surfactants were eliminated as possible candidates, as they tend to inhibit microbial growth and they tend to be retarded in the soil column. Non-ionic, alkyl ethoxylate surfactants, on the other hand, were considered viable candidates for the experiment, as their paraffinic nature is similar to that of diesel fuel, and they are biodegradable in the environment.

An additional criterion that was used as a preliminary screening tool was the ability of the surfactant to form an emulsion. The Hydrophile-Lipophile Balance (HLB) test was used to measure the ability of the surfactant to form an emulsion between water and hydrocarbons. Generally, the greater the HLB, the greater the ability of the surfactant to emulsify hydrocarbons. Based on the results of the preliminary screening, twelve non-ionic surfactants, that had HLB measurements in the range of 8.3 to 14.7, were selected for laboratory testing. This range in HLB measurements was judged to be the best for producing a stable diesel fuel emulsion.

The final screening criteria were based on the performance of the surfactants in a series of experiments. These experiments were used to test twelve non-ionic surfactants (Table 12-2) that were selected during the first screening process for the following properties:

- the biocompatibility of the surfactant with the site specific microbes,

- the ability of the surfactant to emulsify diesel, and

- the foaming properties of the surfactant.

Biocompatibility of the surfactants with the site specific microbes was obviously a critical screening criterion as the surfactant could not be toxic to the site specific microbes. The ability of the surfactants to form a stable diesel fuel emulsion was also considered an important screening criterion, because a stable emulsion would increase the availability of the hydrocarbon compounds for oxidation by the microbes. Foam

Table 12-2   Results of Surfactant Emulsion Experiments (Thin Layer Chromatography Experiment)

Surfactant Name	Manufacturer	Hydrophile-Lipophile Balance (HLB)	Weight (%)	Height Of			Ratio Of		
				Water H_w (cm)	Surfactant H_s (cm)	Diesel H_d (cm)	H_s to H_w	H_d to H_w	H_d to H_s
DI water	Experimental Control	NA	NA	18.0	NA	1.0	NA	0.06	NA
Triton X-100	NR	NA	0.25	16.5	5.3	2.5	0.32	0.15	0.47
Avenel S-150	PPG Industries Inc.	15.4	0.25	16.3	5.0	1.5	0.31	0.09	0.30
Tergitol TMN 10	Union Carbide	16.1	0.25	15.8	4.7	1.5	0.30	0.09	0.32
Tergitol TMN 10	Union Carbide	16.1	0.50	16.7	4.2	1.6	0.25	0.10	0.38
Tergitol TMN 6	Union Carbide	11.7	0.25	17.7	5.5	1.3	0.31	0.07	0.24
Tergitol TMN 10	Union Carbide	16.1	1.00	15.6	NA	2.0	NA	0.13	NA
Neodol 91-6	Shell Chemical Co.	12.5	0.25	15.0	4.8	2.0	0.32	0.13	0.42
Neodol 23-6.5	Shell Chemical Co.	12.5	0.25	16.4	5.4	2.4	0.33	0.15	0.44
Neodol 91-6	Shell Chemical Co.	12.5	NA	17.4	5.2	2.0	0.30	0.11	0.38
Tergitol 15-S-12	Union Carbide	14.7	0.25	14.0	11.0	2.3	0.79	0.16	0.21
Tergitol 15-S-7	Union Carbide	12.1	0.25	12.5	10.0	2.7	0.80	0.22	0.27
Tergitol 15-S-5	Union Carbide	10.5	0.25	16.4	14.0	5.5	0.85	0.34	0.39

Notes:
NR = Not reported
NA = Not applicable
H_w = Measured travel distance of water relative to base of thin layer chromatography plate
H_s = Measured travel distance of surfactant relative to base of thin layer chromatography plate
H_d = Measured travel distance of diesel fuel relative to base of thin layer chromatography plate, normalized to DI water

stability was the final screening criterion examined, because injecting the surfactant in a foam state into the soil would assist in an even delivery of air flow. Foam stability relates the initial foam volume that the surfactant produces to the rate at which the foam volume dissipates. By injecting the surfactant in a foam state, the air flow within the porous media would be less likely to channelize, thus avoiding preferential airflow.

Surfactant Biocompatibility

To assess the biocompatibility of the available surfactants with the site specific microbes, respirometer tests, were conducted that introduced each surfactant into a soil sample obtained during the initial characterization of the study area. The twelve experiments were conducted for a period of 240 hours and oxygen consumption by the microbes was monitored in order to assess whether the surfactant inhibited or promoted the growth of the microbe populations. In all cases, the non-ionic surfactants did not increase or decrease the observed oxygen consumption indicating that the surfactants would not be detrimental to the site specific microbe populations.

Diesel Emulsion

Thin layer chromatography (TLC) was used as a method of testing the ability of the surfactants to emulsify diesel fuel. TLC was also used to assess the potential migration of the surfactants in a porous media. TLC uses a glass plate with small silica particles bound to the surface to simulate a porous medium. In the laboratory experiments, a thin layer of diesel fuel was applied to the bottom two centimeters of a glass plate. The plate was then placed upright into a surfactant and water mixture, and the mixture was allowed to migrate upward, due to capillary forces, for a period of one hour. As the surfactant and water mixture migrates past the thin layer of diesel fuel, the surfactant forms an emulsion with the diesel fuel that will then also migrate upward due to capillary forces.

After a period of one hour, the plates were examined under ultraviolet light to assess the mobility of the surfactant and its ability to emulsify diesel. Under ultraviolet light, the capillary height of the three phases (water, diesel, and surfactant) could be measured (Table 12-2). The heights of the various phases were then used to assess which surfactant would emulsify the diesel fuel the most effectively.

Approximately 25% of the surfactants produced a capillary height of diesel that was less than one centimeter beyond the corresponding height produced by distilled water (used as an experimental control), indicating that these surfactants were least effective in emulsifying diesel (Table 12-2). The majority of the remaining surfactants all produced a height of diesel that was between two and three centimeters beyond that produced by distilled water. The Tergitol 15-S-5 surfactant produced a capillary height of diesel that was 5.5 centimeters high, indicating that this surfactant was the most effective at emulsifying diesel. However, visual inspection indicated that the Tergitol 15-S-5 produced an unstable hydrocarbon emulsion indicating that the diesel fuel and the surfactant would separate as immiscible fluids, thus limiting its' effectiveness.

Foaming Stability

To test the foam stability of the various surfactants, the surfactant was mixed with water and 20 milliliters of solution was placed into a graduated cylinder. The graduated cylinder was then sealed and shaken for approximately two minutes in order to produce an initial volume of foam. The subsequent volumes of foam were then measured in the graduated cylinder at approximately five minute intervals, as well as a final measurement after 200 minutes, to assess the rate at which the foam dissipated. If the rate at which the foam volume decreased was relatively minimal, then the surfactant was judged to be stable. The criterion that was used was whether approximately one half of the initial foam volume remained at the conclusion of the test (200 minutes).

Surfactants produced by the same manufacturer all had similar foam stabilities, with very little difference between the initial volume of foam produced or the rate at which the foam dissipated. For example, all of the Tergitol 15-S series of surfactants initially produced between 40 and 50 milliliters of foam that was judged stable throughout the 200 minute testing period. The Neodal series of surfactants all produced significant volumes of foam initially (e.g., Neodal 91-6 produced over 80 milliliters of foam), but the volume of foam decreased rapidly with time (Figure 12-5).

Final Surfactant Selection
All twelve of the surfactants that were tested in the laboratory did not affect the site specific microbe populations, and therefore biocompatibility was discarded as a selection criterion. The Tergitol 15-S series of surfactants exhibited superior foaming properties, in that they produced a reasonable volume of foam that remained stable during the testing period. The Tergitol 15-S series of surfactants were also considered superior to the other surfactants because they exhibited an above average ability to emulsify diesel fuel.

From the Tergitol 15-S series of surfactants, Tergitol 15-S-12 was selected for field operations. The final decision was based on the superior performance of these surfactants during the laboratory testing, as well as the fact that Tergitol 15-S-12 had a HLB of 14.7. This large HLB indicated that Tergitol 15-S-12 would produce stable water and hydrocarbon emulsions and also provide the best hydrophobic properties, indicating that a stable mixture of Tergitol 15-S-12 in water could easily be maintained.

VAPOR EXTRACTION TEST AND VAPOR EXTRACTION SYSTEM DESIGN

A short term vapor extraction/injection test was conducted in June 1991 to estimate the final areal dimensions of the individual test plots and to calculate the locations of the various vapor extraction wells. The principal purpose of the vapor extraction test was to assess the amount of vacuum that would be produced in the gravelly sand units by the extraction/injection blowers. Several shallow piezometers were drilled at various locations throughout the study area for the purpose of measuring changes in air pressures during the vapor extraction test. A single well was then used for both injection and extraction of air from the gravelly sand units. During the pilot test, both the pressure drops, or the pressure increases for the case of injection, were measured in the piezometers.

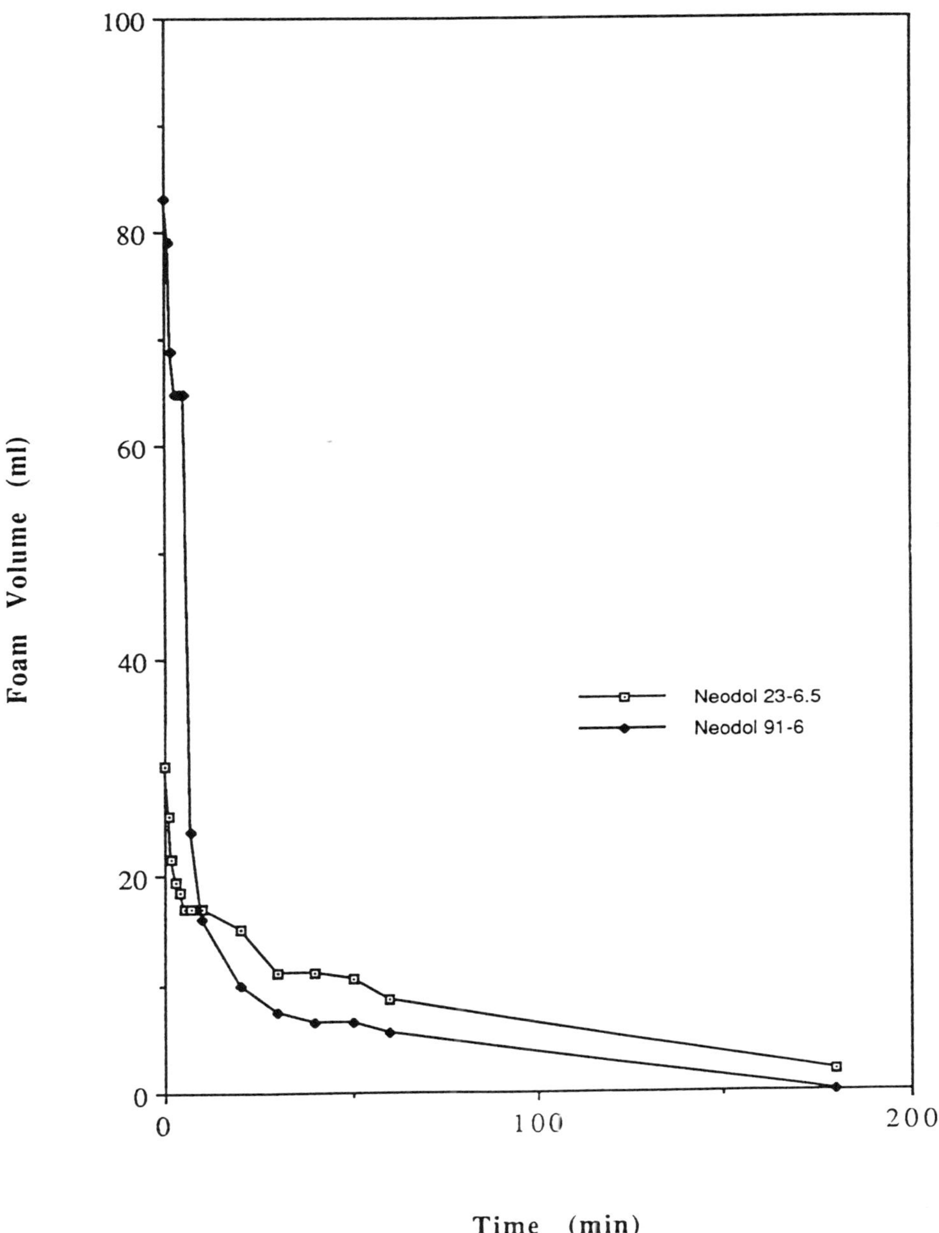

Figure 12-5   Foaming Volume for Neodol Surfactants

Based on the results of the short term tests, it was calculated that one injection well and four extraction wells located on 15 foot centers, would provide the necessary air flow in each of the three bioremediation test plots. Two blowers, one for all of the injection wells and the other for all of the extraction wells, were manifolded to the respective injection or extraction wells using standard polyvinyl chloride (PVC) plumbing fittings (Figure 12-6). Air lines, consisting of standard polybutyl flexible piping, were then installed below grade to connect the individual wells to the manifold. Air flow rates for individual extraction and injection wells were controlled by valves at the manifold. Additionally, due to the size of the blowers, it was necessary to install vents in the manifolds to prevent back pressurization, and subsequent overheating of the blowers.

The calculated upper limit of oxygen consumption by the microbial populations was less than the maximum air flow of 18 cubic feet per minute (cfm) that was derived from the results of the vapor extraction pilot test. Consequently, the flow rates for the injection wells were reduced to six cubic feet per minute (cfm) for the test plots, and STEX extraction wells were controlled to allow an air flow rate of 1.5 cfm. This configuration produced a total air flow rate of six cfm in the four extraction wells, and would provide the daily requirements of oxygen to the microbes.

The delivery of fluids to the nutrient and surfactant test plots was accomplished using a gravity feed injection system. Two 55 gallon drums were used as reservoir tanks, that were then connected to the respective test plots using flexible lines (rubber hoses). Flow rates from the reservoir drums to the flexible fluid lines were controlled by battery operated timers attached to valves at the base of each drum (Figure 12-6). The water supply to the reservoir drums was initially controlled by float valves that were connected to the municipal water supply. However, several failures of the float valves necessitated that the use of the float valves be discontinued, and that the drums be manually filled during weekly maintenance operations.

Initially, all fluids were injected directly into the top of the respective injection wells, with the surfactant mixing with the air stream via an in line mixer placed within the upper two feet of the well in order to produce a foam phase (Figure 12-6). However, due to the relatively high vertical permeability of the soils, the injected supply of nutrients immediately percolated downward beneath the bottom of the well screen. In order to deliver the appropriate amount of nutrients to the soils, it was decided to inject the nutrients along the top of the zone of treatment using two drive points. By injecting into the drive points placed along the top of the zone of treatment the same vertical percolation would produce the necessary downward flow of nutrients through the zone of treatment.

SAMPLING AND ANALYTICAL METHODS

Eight sets of soil samples were collected during the two years of field operations in order to assess the changes in the hydrocarbon concentrations. During 1991, two rounds of soil samples were collected from all four test plots, an initial round that was used for the initial study area characterization and a final round in October 1991. Six sets of soil samples were collected in 1992 from the nutrient and surfactant injection test plots, at approximately four to six week intervals (Figure 12-3). The sampling and analytical methods used to collect and analyze the samples are presented in the following sections.

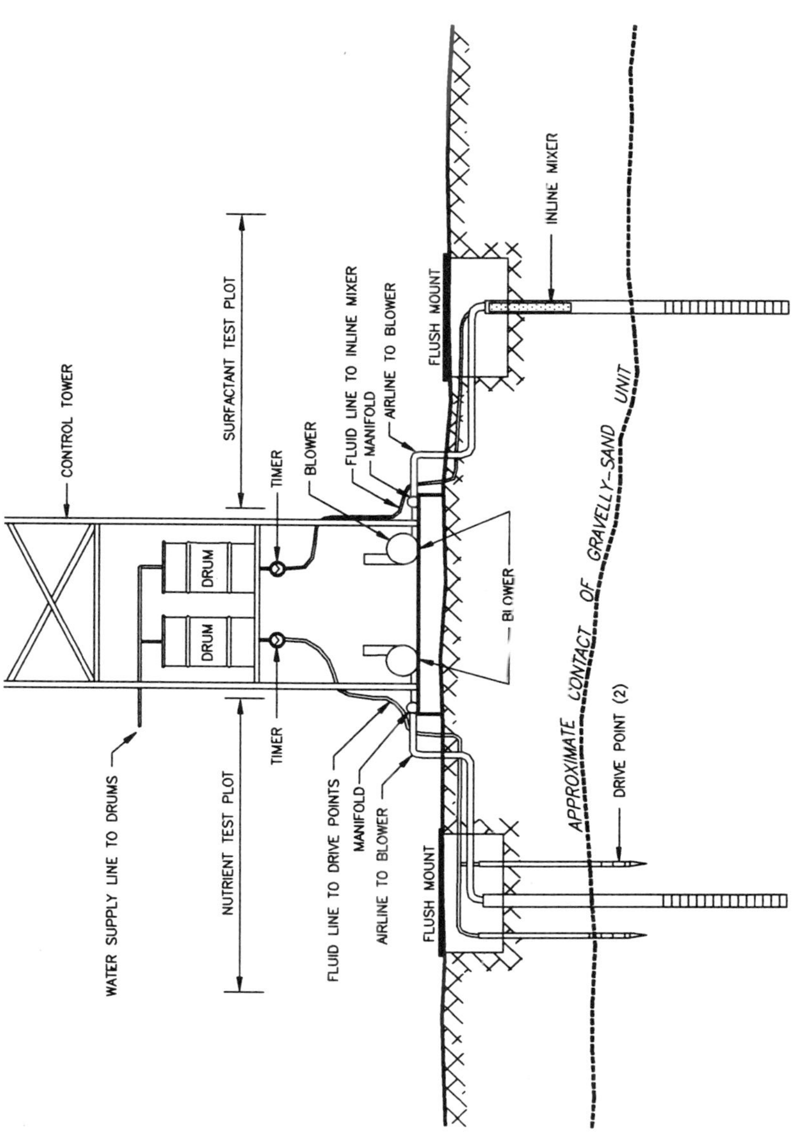

Figure 12-6   Schematic Diagram of Fluid and Air Delivery System

1991 Sampling Methods

All soil samples obtained in 1991 were collected on a regular grid, such that each test plot was sampled at the same relative location in both the horizontal and vertical directions, during both sampling rounds. Each 20 foot by 20 foot square test plot was subdivided into four uniform five foot by five foot square grids. Samples were then numbered using a four digit code that corresponded to:

- test plot location,

- location along the east west grid lines,

- location along the north south grid lines, and

- approximate sample depth.

All of the soils samples were obtained using a 2 inch split spoon sampler that was manually driven with a slide hammer. Two different depths, nominally 8 and 12 feet below land surface, were sampled in order to assess the vertical extent of the total petroleum hydrocarbon (TPH) concentrations in the soils. Minor adjustments to the sampling location were made whenever the sampler could not be driven by hand. In these cases, the soil samples were generally obtained within 1.5 feet of the desired sampling location.

1991 Analytical Methods

Chemical analyses were performed on the soil samples collected in 1991 to evaluate the concentrations of TPH, and to identify the relative amounts of the various molecular compounds comprising the hydrocarbons. These analyses were performed using gas chromatograph (GC) and gas chromatograph/mass spectrograph (GC/MS) methods. For both methods, the hydrocarbons were extracted from the soil samples through a soxoholet extraction. The TPH analyses were completed using a modification of the EPA SW-846 method 8015 with a Hewlett Packard 5890 gas chromatograph with flame ionization detection (GC/FID). The hydrocarbon component analyses were completed using a Hewlett Packard 5970 GC/MS.

1992 Sampling Methods

During the second year of operation the sampling plan was modified in order to obtain a larger number of TPH concentration measurements so that statistical analysis could be made. Four sampling locations were established in each of the two remaining test plots by burying open bottomed 35 gallon drum bottoms within the ballast. For each test plot, a drum bottom was located half way between the injection well and each of four extraction wells, and was subsequently sealed by bolting on a drum lid. The use of the drums allowed for consistent location of the sampling points, thus minimizing the time required to re establish the sampling locations between each sampling event. The four drums in each test plot were sequentially numbered in a counter clockwise direction beginning with the northeast drum. Samples collected during 1992 were identified by a three digit identification number that consisted of:

- the test plot number,

- the drum number, and

- the depth at which the soil sample was collected.

Soil samples were collected by first auguring a borehole using a three horse power hammer drill, and then collecting the soil samples with a 7/16-inch diameter probe. A six to fifteen gram sample of soil was then obtained at nominal depths of eight and ten feet below land surface using the probe. Each sample was split in the field into a top and bottom sample based on the vertical location within the sample probe. In contrast, during the first year of operation, one sample was collected at eight and twelve feet below ground surface. This modification of the sampling plan produced a total of 16 samples for each test plot (per sampling event). Additionally, during the second year of operation, six complete sampling rounds were conducted as a consequence of the longer period of field operations.

1992 Analytical Methods

The 1992 soil samples were analyzed for TPH concentrations using a Hewlett Packard 5890 GC/FID. A modified EPA SW-846 method 8015 analytical procedure was followed for the analysis.

As an additional laboratory quality assurance process, equipment calibration were also conducted approximately once per every eight samples (a minimum of twice daily). This measure was instigated in order to obtain sufficient laboratory control in order to assess variations in the TPH concentrations that were the result of analytical methods.

ANALYTICAL RESULTS

The analytical results of both the 1991 and 1992 sampling events are presented in the following sections. A comparison between the results of the two years was not possible due to problems associated with the GC/FID analyses that prevented an absolute determination of the TPH concentrations. These problems, however, did not affect the comparison of analytical results within 1991, because duplicate measurements of TPH concentrations indicated that the results were consistent, but not representative of the true TPH concentrations. The results obtained during 1991 were therefore used to evaluate relative changes in the measured TPH concentrations and to indicate the relative performance of the different bioremediation methods during the first few months of operation.

Analytical Results From The 1991 Sampling Events

During 1991, all analytical measurements of the TPH concentrations were conducted by CSM. A summary of the analytical results obtained by the GC/FID and GC/MS methods are presented in the following two sections.

Gas Chromatograph/Flame Ionization Detection Analytical Results

The analytical results for the first and second rounds of the 1991 sampling events are presented in Table 12-3. The TPH concentrations for all four test plots did not exhibit any consistent trends with depth. For example, the measured TPH concentrations for

Test Plot 3 (the nutrient only plot) at the 12 foot depth increased between the first and second sampling round at three of the four sampling locations.

Comparison of the percent decrease in the measured TPH concentrations indicates that for all of the test plots, an average reduction of 35% in the TPH concentrations was observed between sampling rounds. There were three exceptions to this observation and they were Test Plot 1 (8-foot samples), Test Plot 2 (8-foot samples), and Test Plot 3 (12-foot samples). In fact, for the individual sample locations, a decrease in the TPH concentrations was observed for 16 out of a total of 25 samples where TPH was detected. However, the computed standard deviations for the reported TPH concentrations equalled or exceeded one half the mean value for both rounds of samples. This indicates that the measured concentrations exhibited considerable variation. These results, while encouraging, were not conclusive.

Gas Chromatograph/Mass Spectrometer Analytical Results
The GC/FID results discussed in the previous section only provided TPH concentrations. To identify which groups of molecular compounds were degrading, several samples were also analyzed using a GC/MS technique. A qualitative comparison of the GC/FID chromatogram from the first and second sampling events in 1991 indicated that several of the peaks representing the smaller (lower weight) molecular compounds had decreased in area.

The chromatogram for a water solution that was ten percent fresh diesel fuel was obtained as a standard, and the chromatograms for soil samples from the same sampling locations obtained during the first and second sampling rounds are shown on Figure 12-7. Each peak of the chromatogram represents a different molecular compound, with the area of the peak being proportional to the concentration of that compound. These samples were analyzed using a ramped column temperature program to provide greater separation between the peaks of the various hydrocarbon components.

The individual peaks on the chromatogram for the fresh diesel fuel sample occurred at regularly spaced intervals indicating a homologous mixture of molecular compounds. In contrast, the peaks on chromatograms for the first and second round of soil samples were irregularly spaced, and individual peaks indicated much greater variation in concentrations. Additionally, the chromatogram for the first and second round of samples had a number of additional peaks appearing between the position of the peaks in the fresh diesel sample. These extra peaks may represent either compounds that are not readily biodegraded by the microbes, or secondary chemicals that form as a result of degradation. As expected, the most easily degradable compounds exhibited substantial decreases in the peak magnitudes, relative to the degradation resistant naphthalene compounds, as observed in the results of both the first and second round of samples (Figure 12-7).

Post-Treatment Microbial Populations
Samples from the second sampling round in 1991 were analyzed to examine post-treatment microbial populations. Respirometer experiments were conducted and the resulting measurements of the microbial populations are presented on Table 12-1.

A comparison of the pre and post treatment CFU (Table 12-1) indicates that an increase in microbial populations was produced in the three enhanced bioremediation

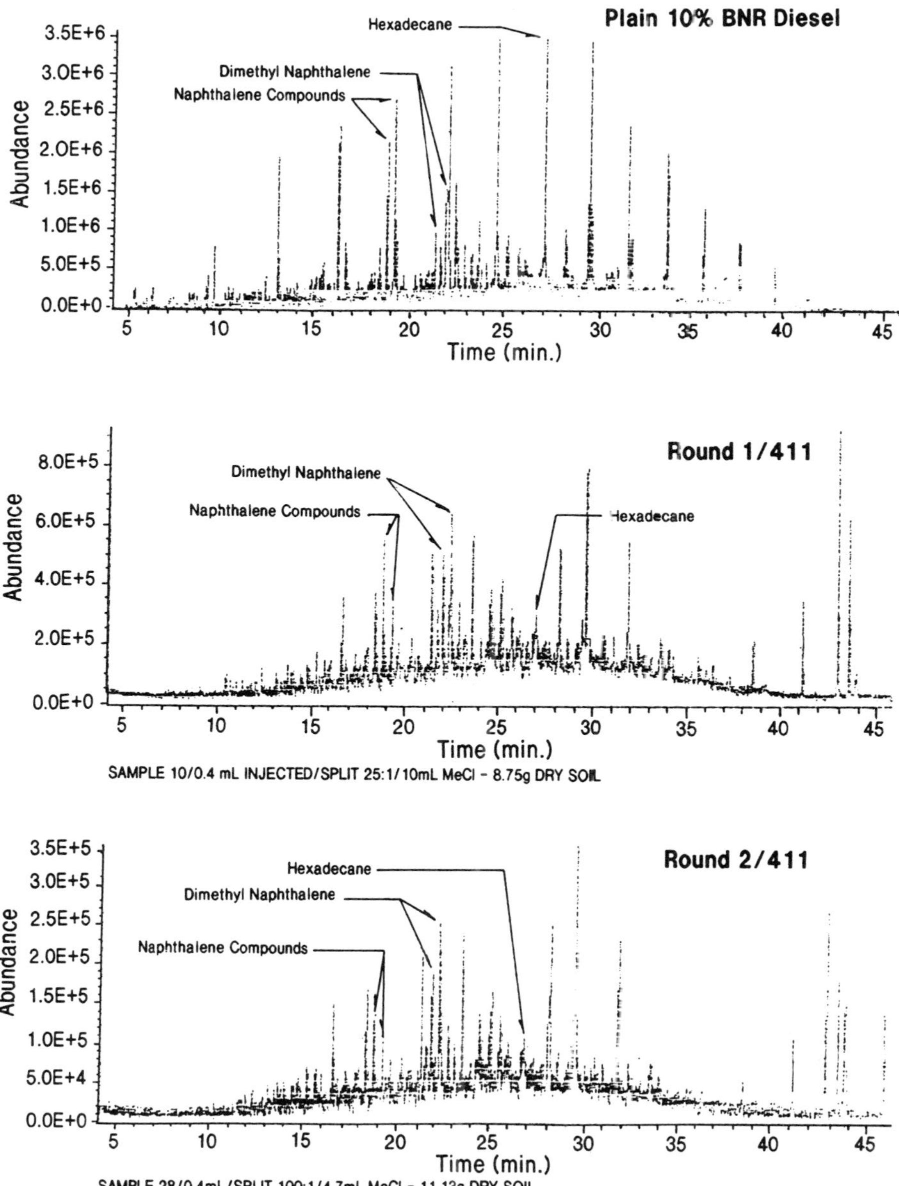

Figure 12-7   A Comparison of Gas Chromatograph/Mass Spectrometer Results from the 1991 Sampling Events

Table 12-3   Summary of Total Petroleum Hydrocarbon Concentrations from 1991 Sampling Events in all Test Plots

Test Plot	1991 Sampling Round	TPH Concentrations - 8-ft samples (g/kg dry soil)				Average	Standard Deviation
		Sample Location 1	Sample Location 2	Sample Location 3	Sample Location 4		
1 Control	1	(0.9)	23.8	7.8	1.9	11.2	11.3
	2	(0.4)	21.8	2.1	4.2	9.4	10.8
2 Surfact	1	(0.8)[1]	1.4	2.7	(0.9)[1]	1.5	0.9
	2	(0.4)[1]	(0.5)[1]	(2.9)[1]	3.0	1.7	1.4
3 Nutrient	1	6.3	6.7	2.6	(0.7)[1]	4.1	2.9
	2	3.4	4.3	1.9	(0.4)[1]	2.5	1.7
4 Air	1	9.8	5.3	3.2	(0.5)[1]	4.7	3.9
	2	2.3	5.8	1.2	(0.4)[1]	2.4	2.4
1	Percent	—	8.40	73.08	-121.05[2]	-13.19[2]	98.85
2	Decrease	—	64.29[3]	-7.41[3]	-233.33[2,3]	-58.82[2]	155.33
3	between	46.03	35.82	26.92	—	36.26	9.56
4	Sampling Rounds	76.53	-9.43[2]	62.50	—	43.20	46.12

Notes:

1. Values shown in parentheses are the analytical detection limit.
2. Negative values indicate an increase in concentration.
3. Changes with nondetect values calculated using the detection limit.

Table 12-3   (Cont.)

Test Plot Round	1991 Sampling	TPH Concentrations - 12-ft samples (g/kg dry soil)				Average	Standard Deviation
		Sample Location 1	Sample Location 2	Sample Location 3	Sample Location 4		
1 Control	1	1.3	3.5	6.8	3.1	3.7	2.3
	2	(0.6)[1]	3.5	(0.7)[1]	(0.6)[1]	1.4	1.4
2 Surfact	1	3.4	4.8	(0.8)[1]	—	3.0	2.0
	2	3.8	(0.4)[1]	(0.5)[1]	2.3	1.8	1.6
3 Nutrient	1	(0.7)[1]	2.6	2.6	—	4.1	2.9
	2	3.5	6.6	1.5	2.0	—	—
4 Air	1	13.5	2.8	(0.9)[1]	4.7	5.5	5.6
	2	3.5	2.4	5.3	1.5	3.2	1.6
1	Percent	53.85[3]	0.00	89.71[3]	77.42[3]	55.25	39.72
2	Decrease	-11.76[2]	91.67[3]	—	—	39.96	73.14
3	between	-400.00[2,3]	-153.85[2]	42.31	—	-170.51[2]	221.63
4	Sampling Rounds	74.07	14.29	488.89[2,3]	68.09	52.15	32.92

Notes:
1. Values shown in parentheses are the analytical detection limit.
2. Negative values indicate an increase in concentration.
3. Changes with nondetect values calculated using the detection limit.

test plots. This enhancement was observed for samples collected at depths between 7 and 14 feet. The control plot (Test Plot 1), on the other hand, did not produce an increase in the microbial populations, and in fact there was a decrease in the number of observed CFU in the same depth range.

The largest increases in microbial populations were observed in the vapor only (Test Plot 4) and nutrient (Test Plot 3) test plots for samples collected at a depth of 12 feet. For the surfactant test plot (Test Plot 2), the increase in the microbial populations was greatest at a depth of eight feet. This result may indicate that the surfactant, in a foam state, was in fact preventing preferential air flow from occurring at the 12 foot depth and in turn produced an increase in the microbe populations at the eight foot depth.

Analytical Results From The 1992 Sampling Events
Tables 12-4 and 12-5 summarize the measured TPH concentrations for all six 1992 sampling rounds for the nutrient (Test Plot 3) and surfactant (Test Plot 4) test plots. As was described earlier, construction of a pipeline through the study area required conversion of the air only test plot (Test Plot 4) into a surfactant test plot. Tables 12-4 and 12-5 also summarize all of the TPH concentrations measured for field and laboratory duplicate samples. A comparison of the spatial distributions of TPH concentrations within a test plot indicate that significant concentration variations were present even at the small scale of the 20 foot square area of the test plots (e.g., the TPH concentrations for soil samples collected in May at the eight foot sampling depth in Test Plot 3 vary from 546 to 8,772 ppm). The analytical results do indicate a general decrease in the TPH concentrations with time. For example, the observed TPH concentrations for Test Plot 4, sample location 3, for soil samples collected at a depth of ten feet (bottom sample split) decreased from approximately 4,560 ppm to 1,690 ppm between May and November 1992.

In several cases, however, the measured TPH concentrations increased in subsequent sampling events. These increases were most prevalent in the October and November 1992 analytical results. For example, approximately one half of the November samples exhibit higher TPH concentrations than those measured in October, despite the fact that the samples were collected from the same approximate depth. Visual inspection of several of the soil samples obtained during this time frame indicated that these samples appeared to contain a higher fraction of silt and clay sized particles than the previous samples, thus, the increasing concentrations may have been the result of preferential sampling of the finer grained fraction that would have higher TPH concentrations.

To assess the potential bias that preferential sampling of the fine grained soils may have had on the measurement of TPH concentrations, several of the soil samples were submitted for grain size analyses. The results of the grain size analyses were then correlated with the measured TPH concentration, using multiple linear regression techniques. The purpose was to evaluate the effect that the various grain size fractions may have on the analytical results. The results of such a correlation could be used to identify whether samples, with smaller grain sized fractions, exhibited higher measured TPH concentrations. If a relationship could be developed, then the multiple linear

Table 12-4 Summary of Total Petroleum Hydrocarbon Concentrations from 1992 Sampling Events in the Nutrient Test Plot (plot 3) (ppm)

Date	Location 3-1				Location 3-2				Location 3-3				Location 3-4			
	8T[1]	8B[1]	10T[1]	10B[1]	8T[1]	8B[1]	10T[1]	10B[1]	8T[1]	8B[1]	10T[1]	10B[1]	8T[1]	8B[1]	10T[1]	10B[1]
5/27/92	1,146	1,996	1,535	4,307	2,130	546	1,669	900	4,663	4,764 4,412L	5,447	1,687	2,609	8,772	3,496	1,654
7/22/92	2,023	2,710	3,834	6,467	663 620L	484	785	1,806	4,102 7,116F	7,331	7,559	3,249	8,335	8,529	3,660 3,933L 3,618F	4,751
9/9/92	346	3,185 1,986L	2,509 2,434F	4,580	5,324 6,120L	867 603L	373	166			1,895	866	1,361	128	1,619 2,045L	1,420
9/30/92	7,991	2,135	4,499	5,677	4,003	76 45L 529F	2,364	743	2,021 1,407F	850	3,339 2,419L	3,197	1,373 488L	669	1,921	1,535
10/22/92	10,685	11,305 10,990L	2,419	2,263	68	44	95	55	2,197	2,005 1,934F	4,099	3,474		2,064 2,504L	6,487 6,874L	2,652 2,403L
11/13/92	1,417	1,961	3,756 2,787F 7,839L	1,007	117	98 85L	63	567	2,575F	3,125 2,912L	4,572 4,206L	2,600	356	1,002	4,320 4,630L	3,961 4,679L 4,742F

Notes:

L = laboratory duplicates

F = field duplicates

1 Each location was sampled at depths of 8 ft and 10 ft, and were subsequently split into two individual samples based on location in the sampling tube: a top (T) split and a bottom (B) split.

Table 12-5 Summary of Total Petroleum Hydrocarbon Concentrations from 1992 Sampling Events in the Surfactant Test Plot (plot 4) (ppm)

Date	Location 4-1		Location 4-2		Location 4-3		Location 4-4									
	8T[1]	8B[1]	10T[1]	10B[1]	8T[1]	8B[1]	10T[1]	10B[1]	8T[1]	8B[1]	10T[1]	10B[1]	8T[1]	8B[1]	10T[1]	10B[1]
5/27/92	1,672	6,925	4,836 8,084F	3,385 5,762F	720	1,453	1,501	4,375	3,815	2,641	3,714	4,558	4,628 4,688F	4,069 7,449F	5,715 5,429F	4,703 3,856F
7/22/92	852	3,287	6,122	7,806	3,792	152 42F	524	1,023	1,865 1,849L	2,103	974 1,449F	2,160	3,547	6,860	5,673 5,733L	6,912
9/9/92	250	777	5,698 5,390L	5,862	39 103F	77 103F	873	204 701F	4,377	4,324	1,879 1,524L	2,081	2,913	659 420L	5,674	5,413
9/30/92	65	1,946	3,955	6,035	10,184	1,428	4,702 5,405F	2,224 467L 3,642F	1,734	934	1,151	2,988	2,636 988L	1,321	1,793	3,749
10/22/92	184	62	6,306	4,003	467	1,716	39	97	2,363	752 821L	2,108	1,867	6,038 6,151L	1,440	3,956 3,361L	5,419 4,816L 9,579F
11/13/92	5,865	6,839	5,966	6,263 6,173L	15,500 13,867L	15,233 15,133L	2,068	4,303 3,791F	5,304 2,080L	1,482 1,255L	2,190 1,836F	1,693	4,018	1,175	4,745 5,875L	6,402 6,491L

Notes:

L = laboratory duplicates

F = field duplicates

[1] Each location was sampled at depths of 8ft and 10ft, and were subsequently split into two individual samples based on location in the sampling tube: a top (T) split and a bottom (B) split.

regression procedures could be utilized to remove the influence of the smaller grain size fractions on the observed TPH concentrations.

The calculated correlation coefficient for a multiple linear regression of all the grain size fractions with the corresponding TPH concentration was -0.54. This low correlation coefficient was attributed to the relatively small mass of the individual soil samples available to perform the analyses (approximately one tenth the mass required by the ASTM standard). The limited sample mass probably did not provide a representative subsample for each soil sample. Thus, there may have been insufficient mass to obtain a valid correlation between the grain size distributions and the TPH concentrations in the samples (theoretically, the correlation coefficient should have been -1.0). The calculated correlation coefficient for the smaller grain sized fractions (silt size and smaller) with the TPH concentrations was significantly lower and it was calculated to -0.40. Because the correlation coefficients were so low no attempt was made to correct the measured TPH concentrations based on the grain size distribution of the soil samples.

STATISTICAL ANALYSES

Statistical analyses were conducted using the TPH concentrations obtained during the 1992 field season to estimate the effects of both nutrient and surfactant injection on the diesel bioremediation processes. Due to the relatively large variations in the measured TPH concentrations, advanced statistical methods were used. This section summarizes the general character of the TPH concentration data, the accuracy of the laboratory analyses, and the various statistical methods that were used to analyze the data.

Preliminary Review Of 1992 TPH Concentration Data

An initial review of the 1992 analytical results indicated that the measured TPH concentrations varied considerably between sampling events. In some cases, increases in the TPH concentrations were observed, while in other cases substantial decreases in the TPH concentrations were measured (Tables 12-4 and 12-5). The variations in the TPH concentrations were attributed to laboratory analytical accuracy and the heterogeneity of soils at the study area.

The contribution of laboratory analytical accuracy to the TPH concentration variations was analyzed and found to be insignificant. To quantitatively evaluate the effects of laboratory analytical accuracy on the resulting data, all the laboratory duplicate samples were analyzed for:

- the relative percent error (the ratio of the difference between the laboratory duplicate result and the analyzed sample result, to the analyzed sample result) for each pair of laboratory duplicates, and

- the frequency of relative percent error among all the duplicate samples.

Table 12-6 summarizes the results of the relative percent error and the frequency of the relative percent error for the 36 pairs of the laboratory duplicates. The mean relative percent error for the laboratory duplicates was 19.2%, with a standard deviation of 20.1%. This high standard deviation was mainly due to three inconsistent measurements that produced relative percent errors above 60% (79%, 60%, and 62% respectively). Frequency analysis shows that 72.2% of the laboratory duplicate samples had a relative error below 25%. Based upon the preliminary review of the data, it was concluded that the contribution of laboratory analysis error to the data variation is insignificant. The primary cause for the data variation, therefore, was probably due to the heterogeneity of the soils in the study area.

The heterogeneity of the soils in the study area was evaluated by examining the TPH concentrations that were measured for the field duplicates. Table 12-6 summarizes the relative percent error and the frequency of the relative percent error for the 21 pairs of the field duplicates. The mean relative percent error for the field duplicates was 79.6%, with a standard deviation of 132.1%. This large standard deviation indicated that the variations in the observed TPH concentrations can primarily be explained by the heterogeneity in the soils between closely spaced sampling locations. Frequency analysis indicates that over 60% of the field duplicate samples had a relative percent error greater than 25%, which further confirms that soil heterogeneity was the primary source of the data variation.

Nonparametric Statistical Analyse
Given the observed variations in the TPH concentrations, several different nonparametric statistical methods were used to assess whether enhanced bioremediation is effective in reducing TPH concentrations. Nonparametric statistical tests do not necessarily require normally distributed data. For example, many nonparametric techniques are based on a relative ranking of the data, where the absolute value or magnitude of the data are considered only to the extent that they are used to order the data from highest to lowest values. Two such nonparametric methods, the Mann-Kendall test for trends[13,14], and the distribution free signed rank test method developed by Wilcoxon[15], were used to analyze the TPH concentration data.

Mann-Kendall Test for Trends
The Mann-Kendall test was used to examine the TPH concentrations to identify decreasing trends for a specified test plot through time. Statistically valid downward trends were interpreted to indicate that enhanced bioremediation was increasing the naturally occurring degradation of the diesel hydrocarbons. For the Mann-Kendall test, a significance level of 0.25 was selected as the minimum significance level for defining a statistically valid result.

The Mann-Kendall method involves determining the relative sign for all of the possible differences between the TPH concentrations. For example, a value of 1 was assigned to each positive difference, zero to no difference, and -1 to each negative difference. The relative signs are then used to calculate the Mann-Kendall statistic (S), by subtracting the number of negative differences from the number of positive differences. If the calculated Mann-Kendall statistic (S) is a large negative number, and the probability value (α) corresponding to S is less than the specified significance level, then the hypothesis that a downward trend in the TPH concentrations over time is present (i.e., the TPH concentrations are decreasing through time) can be accepted.

The Mann-Kendall test results are summarized in Table 12-7 and the tabulated results correspond to:

Table 12-6   Relative Percent Error of 1992 Analytical Results for Duplicate Laboratory and Field Samples

Summary of the Relative Error for Laboratory Duplicate Samples

Sample Size	Mean	Standard Deviation	Frequency			
			0-25%	25%-50%	50%-75%	75%-100%
36	19.2%	20.1%	72.7%	16.7%	8.3%	2.8%

Summary of the Relative Error for Field Duplicate Samples

Sample Size	Mean	Standard Deviation	Frequency				
			0-25%	25%-50%	50%-75%	75-100%	>100%
21	79.6%	132.1%	38.1%	14.3%	28.6%	4.8%	14.3%

- the number of observed data (N),

- the calculated Mann-Kendall statistics (S), and

- the probability value corresponding to calculated Mann-Kendall statistic (α).

Based on a significance level of 0.25, 12 out of the 32 sampling locations produced a decreasing trend, with eight of the trends being observed in the nutrient only test plot (Test Plot 3). Interestingly, one half of the 16 sampling locations in the nutrient only test plot produced statistically recognizable decreasing trends in TPH concentrations. This result indicates that the addition of nutrients as an enhanced bioremediation method was superior to use of surfactants, because only four of the sampling locations in the surfactant test plot showed decreasing trends.

Wilcoxon Distribution–Free Signed Rank Test
The Wilcoxon method was originally developed to assess the effects of medical treatments on patients. For this investigation, the analogy incorporated comparing the initial TPH concentrations in the soils (May 1992 samples) to the subsequent TPH concentration data for the purpose of identifying decreases resulting from the use of nutrients and surfactants. The heterogeneity among patients in the treatment would correspond to the observed heterogeneity of the study area. The treatment would correspond to the application of nutrients or surfactants. This statistical method, therefore, was considered ideal for assessing the effects of nutrient and surfactant injection on the bioremediation processes.

The May 1992 data from Test Plots 3 and 4 were used as the pretreatment results, and the subsequent data collected in July, August, September, October, and November, were used as the respective post treatment samples. As described previously, a preliminary review of the measured TPH concentrations identified several samples with anomalous TPH concentrations (3-1-8T (9/30), 3-1-8T (10/22), 3-1-8B (10/22), 4-2-8T (9/30), 4-2-8T (11/13), and 4-2-8B (11/13)). Because these samples were shown to be inconsistent with the remaining TPH concentration data they were omitted from the data set used for the Wilcoxon test.

The test procedure for the Wilcoxon method involves:

- calculating the absolute difference between the observed TPH concentrations for each treatment phase and the pretreatment concentration,

- computing the T^+ statistic (sum of the positive signed ranks) based on the rank order and the sign of the difference, and

- assessing the treatment effect according to a one sided test at a specified level of significance.

Table 12-8 summarizes the results of the Wilcoxon test, and is based on the assumption that the May (1992) TPH concentrations represented the pretreatment samples. In Table 12-8, the computed significance level (α) represents the lowest significance level at which the hypothesis that a positive treatment effect has occurred is valid. Correspondingly, the lower the computed significance level (α) the more pronounced the treatment has influenced the biodegradation of the hydrocarbons.

Table 12-7   Results of the Mann-Kendall Nonparametric Statistical Tests

Sample Split	Test Plot 3 (Nutrient)				Test Plot 4 (Surfactants)			
	Location 3-1	Location 3-2	Location 3-3	Location 3-4	Location 4-1	Location 4-2	Location 4-3	Location 4-4
8 ft Top	$N = 5$ $S = 5.0$ $\alpha = 0.235$ upward trend	$N = 6$ $S = -5$ $\alpha = 0.235$ *downward trend*	$N = 4$ $S = -4$ $\alpha = 0.167$ *downward trend*	$N = 5$ $S = -6$ $\alpha = 0.117$ *downward trend*	$N = 6$ $S = -3$ $\alpha = 0.360$ no trend	$N = 6$ $S = 5.0$ $\alpha = 0.235$ upward trend	$N = 6$ $S = 3.0$ $\alpha = 0.360$ no trend	$N = 6$ $S = -1$ $\alpha = 0.50$ no trend
8 ft Bottom	$N = 6$ $S = 1$ $\alpha = 0.50$ no trend	$N = 6$ $S = -7$ $\alpha = 0.136$ *downward trend*	$N = 5$ $S = -2$ $\alpha = 0.408$ no trend	$N = 6$ $S = -5$ $\alpha = 0.235$ *downward trend*	$N = 6$ $S = -5$ $\alpha = 0.235$ *downward trend*	$N = 6$ $S = 7.0$ $\alpha = 0.136$ upward trend	$N = 6$ $S = -7$ $\alpha = 0.1360$ *downward trend*	$N = 6$ $S = -5$ $\alpha = 0.235$ *downward trend*
10 ft Top	$N = 6$ $S = 3$ $\alpha = 0.360$ no trend	$N = 5$ $S = -4$ $\alpha = 0.242$ *downward trend*	$N = 6$ $S = -1$ $\alpha = 0.500$ no trend	$N = 6$ $S = 5$ $\alpha = 0.235$ upward trend	$N = 6$ $S = 3$ $\alpha = 0.360$ no trend	$N = 6$ $S = 1$ $\alpha = 0.50$ no trend	$N = 6$ $S = 3$ $\alpha = 0.360$ no trend	$N = 6$ $S = -7$ $\alpha = 0.136$ upward trend
10 ft Bottom	$N = 6$ $S = -7$ $\alpha = 0.136$ *downward trend*	$N = 6$ $S = -7$ $\alpha = 0.136$ *downward trend*	$N = 6$ $S = 3$ $\alpha = 0.360$ no trend	$N = 6$ $S = 5$ $\alpha = 0.235$ upward trend	$N = 6$ $S = 3$ $\alpha = 0.360$ no trend	$N = 6$ $S = -3$ $\alpha = 0.360$ no trend	$N = 6$ $S = -11$ $\alpha = 0.028$ *downward trend*	$N = 6$ $S = 3$ $\alpha = 0.360$ no trend

Notes:

N = Number of observations used in analysis

S = Mann-Kendall statistic

α = Probability associated with Mann-Kendall statistic tabulated

upward trend — Indicates that a statistically (significance level = 0.25) valid increase in total petroleum hydrocarbon concentrations was observed

no trend — Indicates that no trend was observed in total petroleum hydrocarbon concentrations

downward trend — Indicates that a statistically (significance level = 0.25) valid decrease in total petroleum hydrocarbon concentrations was observed (italics table entries)

Treatment effects were observed in the nutrient test plot (Test Plot 3), with a minimum of a 0.211 significance level, beginning in September. In fact, with the exception of the October treatment, the significance level associated with the nutrient treatment would be approximately 0.10. For the surfactant test plot (Test Plot 4), statistically valid treatment effects were observed at a significance level greater than 0.05 for three of the five post treatment sampling periods (9/9/92, 9/30/92, 10/22/92).

The relationship of significance level with time can also be used to interpret the effectiveness of the various treatment techniques. A close examination of (α) values with time for Test Plot 3 and Test Plot 4 data (Figure 12-8) indicate that:

- the significance levels for Test Plot 3 exceeded the corresponding values for Test Plot 4 (except the November data), indicating that the surfactant exerts a greater effect on the bioremediation process than nutrients;

- the significance levels for both Test Plot 3 and Test Plot 4 decrease with time suggesting that the difference between pretreatment (May data) and each post treatment TPH concentration increased over time;

- the decreasing trend of significance levels with time suggests that the diesel consumption by the *in situ* microbes increased over time.

In conclusion, the results of the Wilcoxon data analysis suggest that both nutrient and surfactant injection exert a positive effect on the diesel bioremediation process, and that the effect of surfactant is more evident than that of the nutrient only injection.

Summary of the Results of the Statistical Analyses
Two nonparametric statistical tests were conducted to assess the potential effect that the various methods of enhanced bioremediation may have on the biodegradation of diesel fuel. The first statistical test, the Mann-Kendall test for trends, indicated that supplementing the site soils with additional nutrients was preferable to the use of surfactants. The Wilcoxon test, on the other hand, indicated that the use of surfactants as a method of enhancing the naturally occurring biodegradation was marginally better than the method that relied solely on providing the microbes additional nutrients. In general it was concluded that the Wilcoxon test provides a higher degree of confidence in the test results. Correspondingly the use of surfactants provides a higher degree of enhancement of biodegradation. However, both methods were shown to influence the overall degradation of diesel.

CONCLUSIONS AND RECOMMENDATIONS

The results of the two year investigation of methods of enhancing *in situ* bioremediation of diesel hydrocarbons indicate that supplying the naturally occurring microbes with the nutrients that are found to be deficient in site soils, and injection of surfactants into the soils, positively influenced the degradation of subsurface hydrocarbons. Because both of these methods of enhanced bioremediation were shown to be beneficial, the final selection of the preferred method considered other factors.

Supplementing the site soils with deficient nutrients was considered the preferred method of enhancing *in situ* bioremediation because of the smaller material costs, as well as operational considerations that are involved with using surfactants. During the current study, the approximate material costs associated with using nutrients were on the order of $0.30 per cubic yard of treated soil. Conversely, the material costs associated with the use of a surfactant to enhance the bioremediation process were approximately $8.50 per cubic yard of treated soil. These cost estimates were based on the purchase price of the nutrients and the surfactant that were used during the study, and do not include drilling costs, blower costs, or other capital expenditures as these costs were considered equivalent for both methods.

Furthermore, due to the additional engineering requirements associated with the injection of surfactant in a foam state, the long term operation and maintenance costs associated with the use of nutrients as a bioremediation method should be considerably less than those associated with the use of surfactants. A final consideration was based on a regulatory perspective, such as those regulations associated with injecting either nutrients or surfactants into the subsurface. In terms of the current federal regulations,

Table 12-8   Results of Wilcoxon Nonparametric Statistical Test

Case[1]	Nutrient Test Plot 3	Surfactant Test Plot 4
5/27/92 samples versus 7/22/92 samples	N = 16 T = 107 α = 0.5	N = 16 T = 56 α = 0.262
5/27/92 samples versus 9/9/92 samples	*N = 14* *T = 32* α *= 0.108*	*N = 16* *T = 35* α *= 0.041*
5/27/92 samples versus 9/30/92 samples	*N = 15* *T = 45* α *= 0.211*	*N = 15* *T = 30* α *= 0.021*
5/27/92 samples versus 10/22/92 samples	*N = 13* *T = 26* α *= 0.095*	*N = 16* *T = 21* α *= 0.007*
5/27/92 samples versus 11/13/92 samples	*N = 15* *T = 38* α *= 0.094*	N = 14 T = 53 α = 0.5

Notes:

[1] 5/27/92 samples were considered pre-treatment samples. All other samples were considered post-treatment samples.

N = Number of observations in the analyses

T = Computed Wilcoxon T statistic

α = Computed significance level

Italics entries correspond to results that indicate that a statistically valid (significance level of 0.25) reduction in total petroleum hydrocarbon concentrations resulted from the treatment (the use of nutrients or surfactants).

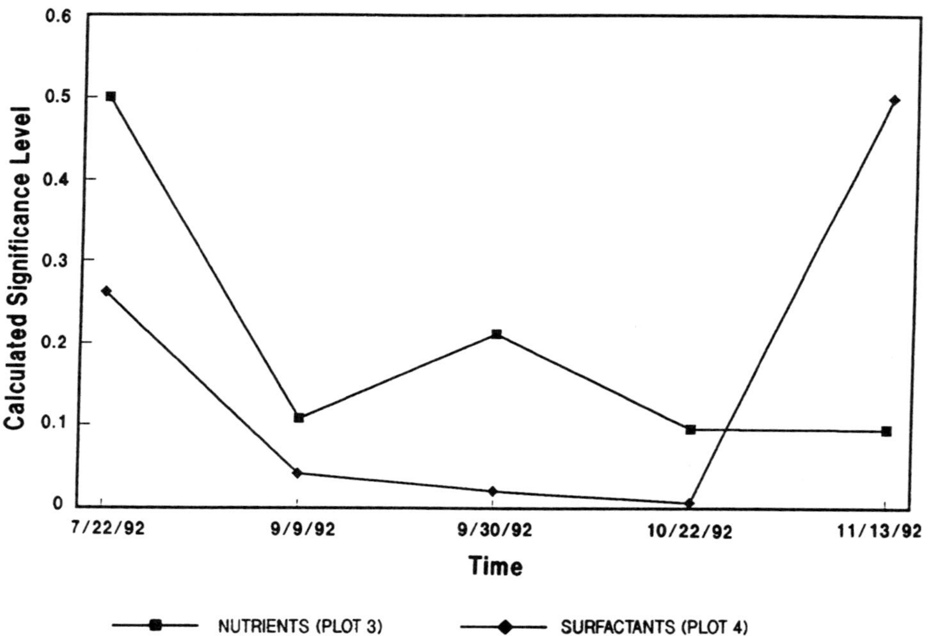

Figure 12-8   Wilcoxon Significance Levels vs. Time for Nutrients and Surfactants

nutrient enhancement is the preferred method as there is a lack of research and data that address potential ancillary effects of surfactant injection. Federal injection regulations may not be applicable to the use of nutrients as a method of enhancing the bioremediation of diesel fuel, if the nutrients were introduced to the subsurface by some type of surface application or a shallow drip irrigation method.

Quantifying the rate at which diesel concentrations were reduced would be an obvious extension of this investigation. Due to the heterogeneity of the study area, which produced considerable variation in the observed total petroleum hydrocarbon concentrations in the soils, the current investigation was unable to quantify how much the various enhanced bioremediation methods increased the overall rate of diesel hydrocarbon remediation. In order to minimize the effects that geologic heterogeneity have on the experimental results, numerous (6 to 12) soil samples would be required in order to obtain a representative total petroleum hydrocarbon concentration for each sample location. In this case, the quantity of samples that would need to be obtained would preclude the use of conventional soil sampling as a sampling method, and alternative sampling methods such as porous cup lysimeters or similar devices, and soil gas sampling methods would need to be considered.

Porous cup lysimeters represent a method of obtaining a sample of *in situ* soil water, and could be used to obtain repeated soil moisture samples from one location. Permanent installation of the lysimeter would minimize the influence of local soil heterogeneities as well as variations in the observed total petroleum hydrocarbon concentrations that are the result of the disruptive effects of conventional soil sampling methods. Another advantage of using a porous cup lysimeter would be that numerous samples could be obtained over a short period of time, thus providing a statistically valid number of samples to characterize the actual subsurface conditions. One disadvantage of a porous cup lysimeter would be that samples cannot be obtained over the entire range of soil moisture contents, because most porous cup lysimeters are limited to approximately -1.0 bar of soil suction. This lower limit of soil suction would limit the application of lysimeters at sites that are known to have either coarse grained sediment or are located in arid climates. The aqueous solubility of diesel range hydrocarbons would also limit the application of porous cup lysimeters in that the fraction of diesel fuel retarded within the soils would not be representatively sampled.

Soil gas sampling, ideally using permanently installed soil probes, represents another method of obtaining *in situ* measurements of total petroleum hydrocarbon concentrations. The use of permanently installed soil probes would, as with porous cup lysimeters, minimize both the influence of geologic heterogeneities and sampling methods on the measured sample concentrations. Soil gas sampling methods would be less influenced by site conditions than porous cup lysimeters, in that they would not be restricted by any soil moisture restrictions. However, a disadvantage of using a soil gas sampling method would be that the resulting measurements of total petroleum concentration would only represent the more volatile fraction of hydrocarbon compounds found in diesel fuel.

Another factor that future investigations need to address is ability of the various enhanced bioremediation methods to degrade subsurface hydrocarbon levels to specified soil cleanup standards. To assess how effective the enhanced bioremediation methods are in achieving soil cleanup standards, it will be necessary to investigate the efficiency of each method when relatively low hydrocarbon concentrations are present in the soils.

The results of the investigation indicated that the use of both nutrients and surfactants increased the bioremediation of diesel hydrocarbons in the soils over vapor extraction alone. Nutrient enhanced bioventing and surfactant enhanced bioventing are viable treatment techniques for sites that are either in remote locations, or sites that require a minimal impact to ongoing site activities. Of these two methods, the use of surfactant enhanced bioventing for enhancing bioremediation was shown to produce a marginally greater effect than use of nutrients for the *in situ* degradation of diesel hydrocarbons in the vadose zone. However, the addition of deficient nutrients alone is a more cost effective method, considering both the costs of treatment and the difficulties associated with injecting the appropriate quantities of surfactants into the vadose zone.

REFERENCES

1.   Kampbell, D., 1991, Bioventing/Biodegradation remediates liquid hydrocarbons in unsaturated zone: EPA/540/M-91/004, no. 6, p. 1-2.

2.   Sterrett, R.J., Ransom, M.E., and Barnhill, G.D., 1985, On-site treatment of a pesticide spill: in Proceedings of Conference on Hazardous Material Spills, May 5-8.

3.   Loehr, R.C., 1991, Adjust key factors to enhance microbial activity: *Soils*, November-December, p. 42-53.

4.   Raymond, R.L., Jamison, V.W., and Hudson, J.O., 1976, Beneficial simulations of bacterial activity in groundwater containing petroleum products: American Institute of Chemical Engineering Symposium Series, Vol. 73, p. 390-404.

5.   Wilson, J., 1991, Nitrate enhanced bioremediation restores fuel contaminated groundwater to drinking water standard: EPA/540/M-91/002, no. 5, p. 1-2.

6.   Currie, J.K., Bunge, A.L., Updegraf, D.M., and Batal, W.H., 1991, Surfactant enhanced remediation of creosote contaminated soils: in Sixth National Conference on Hydrocarbon Contaminated Soils: Analysis, Fate, Environmental & Public Health Effects and Remediation, University of Massachusetts at Amherst, September 23-26.

7.   Aronstein, B.N., Calvillo, Y.M., and Alexander, M., 1991, Effect of surfactants at low concentration on sorbed aromatic compounds in soil: *Environmental Science and Technology*, Vol. 25, no. 10, p. 1723-1736.

8.   West, C.C., and Harwell, J.H., 1992, Surfactants and subsurface remediation: *Environmental Science and Technology*, Vol. 26, no. 12, p. 2324-2330.

9.   Hansen, W.R., and Crosby, E.J., 1982, Environmental geology of the front range urban corridor and vicinity, Colorado: United States Geological Survey Professional Paper 1230, 99 pp.

10. Robson, S.G., and Romero, J.C., 1981, Geologic structure, hydrology and water quality of the Denver aquifer in the Denver basin, Colorado: United States Geological Survey Hydrogeologic Investigation Atlas HA-646.

11. Costa, J.P., and Bilodeau, S.P., 1982, Geology of Denver, Colorado, USA: Bulletin of The Association of Engineering Geologists, Vol. XIX, no. 3, p. 261-314.

12. Hickman, G. T., and Novak, J.T., 1989, Relationship between subsurface biodegradation rates and microbial density: *Environmental Science and Technology*, Vol. 23, no. 5, p. 525-532.

13. Mann, H.B., 1945, Non-parametric tests against trend: *Econometrica*, vol. 13, p. 245-259.

14. Kendall, M.G., 1975, Rank correlation methods: 4th ed. Charles Griffin, London.

15. Hollander, M., and Wolfe, D.A., 1973, *Nonparametric statistical methods:* Wiley, New York.

Questions and Answers:

Q. I have a couple of questions about the holes that you dug for sampling. Were they plugged after you sampled?

A. We used a barrel with a seal over the top to get through the fill material. The barrel was sealed between sampling events to keep from short circuiting through the system.

Q. So short circuiting was not a problem with the samples that you took?

A. That's right and we did plug the tops of each of the boreholes with grout after they were sampled.

Q. My other question was about your plugging of the pores. Do you know if that was from aggregates of your biological organisms or do you think that was from precipitation of your surfactants with your groundwater?

A. I think it's the foam itself. We did not however investigate slime buildup within the porous medium itself.

Q. So you don't know for sure then?

A. No.

Q. During your work did you do a cost comparison between anything besides surfactants and nutrient addition? Was there a baseline cost comparison with land filling? Secondly, was this strictly done as a pilot study or was this going to be expanded? It seems like it was a lot of expense to get into for such a small area. Was it to scale up to a much larger area?

A. The goal was to look at processes and to look at some cost effectiveness, so we did not make a comparison to, say, land filling. It was a comparison between nutrients versus surfactants. It was only performed on a pilot scale, field scale, strictly for research purposes. There was no intention to ramp it up. If you wanted to go to production we certainly have lots of insights and I'll give you several of those. One would be - you certainly want to weatherproof the system. You want to be able to get started at the first sign of spring, especially in the Rocky Mountains, so that you can obviously take advantage of the warm season. Those would be the two critical factors.

Q. I'm sorry, I didn't catch the nature of the nutrients and surfactants that you used in the study.

A. We used diammonium phosphate for the nutrients and the name of the surfactant was Tergitol 15-S-12. It is produced by Union Carbide. It's a non-ionic surfactant.

ACRONYMS

AF	soil-to-skin adher
AFB	Air Force Base
AHH	aryl hydrocarbons hydroxylase
API	American Petroleum Institute
ASTM	American Society for Testing and Material
AT	average time
Avg. wt.	average weight
BGS	below ground surface
BP	boiling point
BTEX	benzene, toluene, ethylbenzene, and xylenes
BW	body weight
C	Celsius
C:N	carbon to nitrogen
CAS	Chemical Abstract Service
CDI	chronic daily intake
CDI_d	chronic daily intake
CF	conversion factor
CFM	cubic feet per minuet
CFU/ml	colony -forming units per milliliter
CI	confidence interval
CLP	coal liquidation product
CNS	central nervous system
COC	contaminants of concern
C_p	leachate concentration
CRWQCB	California Regional Water Quality Control Board
CS	chemical concentration in soils
C_{si}	concentration of species i
CSM	Colorado School of Mines
CSU	Colorado State University
CSXT	CSX Transportation
DA	dermal absorption factor

EAD	equivalent aerodynamic diameter
EAF	environmental attenuation factor
ED	exposure duration
ED	exposure duration
EED	estimated exposure dose
EF	exposure frequency
EF	exposure frequency
EPA	U.S. Environmental Protection Agency
EW	body weight, in kg
F	Fahrenheit
FI	fraction ingested
FI	fraction ingested from source
FID	flame ionization detection
f_{oc}	fraction of organic carbons
g	gravitational acceleration
g/kg	grams per kilogram
g/ml	gram per milliliter
GC	gas chromatography
HBG	health-based cleanup goals
HBGgw	health-based cleanup goal for groundwater
HCI	Hydrolic Consultants, Inc.
HDOH	Hawaii Department of Health
HDPE	high density polyethylene
HDPE	high density polyethylene
HEAST	Health Effects Assessment Summary Tables
HI	hazard index
HLB	hydrophile-lipophile balance
HP	horsepower
HQ	hazard quotient
HRT	hydrocarbon recovery trench
IR	infrared
IR	ingestion rate
IR	inhalation rate
J	hydraulic gradient
k	intrinsic permeability of porous medium
K_d	partition coefficient

k_{Ii}	mass transfer coefficient
K_{oc}	carbon partition coefficient
K_{ow}	octanol water coefficient
L/kg	liters per kilogram
LD_{50}	lethal dose to 50
LNAPL	light nonaqueous phase liquid
LOAEL	lowerst-observed-adverse-effect level
LUFT	leaking underground fuel tank
MCL	maximum contamination level
MF	modifying factors
mg/kg-day	milligrams per kilometer per day
mg/L	milligrams per liter
mg/L	milligrams per liter
mg/m^3	milligrams per cubic meter
mL/kg	milliliters per kilogram
MMAD	mass median aerodynamic diameter
MOE	margin of exposure
MW	molecular weight
Nc	the capillary number
NJDEP	New Jersey Department of Environmental Protection
NOAEL	no-observable-adverse-effect level
$NOAEL_{ADJ}$	adjusted normal-observed-adverse-effect level
NPAH	noncarcinogenic polycyclic aromatic hydrocarbons
NPL	National Priority Listed
NRC	National Research Council
NYSDEC	New York State Department of Environmental Conservation
O	interfacial tension
OSF	on site storage facility
PAH	polynuclear aromatic hydrocarbons
PEF	particulate emission factor
PEF	particulate emission factor
PEIS	primary eye irritation score
PPM	parts per million
PPM	parts per million
PRA	preliminary risk assessment

PTYG	peptone-tryptone-Yeast-Glucose
PVC	polyvinyl chloride
P_w	the density of water
RfD	reference dose
RFD_d	dermal reference dose
RFD_i	inhalation reference dose
RFD_o	oral reference dose
RI/FS	Remedial Investigation/Feasibility Study
RME	reasonable maximum exposure
ROD	record of decision
SA	skin surface area
S_o	organic or liquid contaminant saturation
S_{or}	residual liquid contaminant saturation
SPTCo	Southern Pacific Transportation Company
SVOC	semivolitile organic compounds
TCLP	toxicity characteristic leaching procedure
TEPCO	Tesoro Environmental Products Company
TIC	tentatively identified compounds
TLC	thin layer chromatography
TLV	threshold limit value
TPH	total petroleum hydrocarbons
TPH-diesel	total petroleum hydrocarbons as diesel
TWA	time-weighted average
UF	uncertainty factors
USEPA	United States Environmental Protection Agency
UST	underground storage tank
V_{do}	volume of discontinuous liquid contaminant
VES	vapor extraction system
V_o	volume of hydrocarbons
VOC	volatile organic compounds
V_v	volume of voids
WI	water intake
yd^3	cubic yard

CONTRIBUTORS

DEAN ANSON II, Anson Environmental Ltd., Huntington, NY

JAMES BAKER, Alton Geoscience, Irvine, CA

JEFF BORDELON, Radian Corporation, Milwaukee, WI

JAMES L. BROWN, OHM Remediation Services Corp., Trenton, NJ

MELVIN L. BURDA, Burlington Northern Railroad, Overland Park, KS

DAVID CLARK, The Atchison Topeka and Sata Fe Railway, Topeka, KS

JAMES J.J. CLARK, Alton Geoscience, Irvine, CA

KENT A. FRIESEN, Engineering-Science, Inc. Denver, CO

BOB FRONCZAK, Radian Corporation, Milwaukee, WI

E. KINZIE GORDON, Engineering-Science Inc., Denver, CO

MICHAEL J. GRANT, Southern Pacific Transportation Co., Boise, ID

M. NEAL GUENTZEL, University of Texas, San Antonio, TX

PETER R. GUEST, Engineering-Science, Inc., Denver, CO

MICK HARDIN, The Atchison, Topeka, and Santa Fe Railway Co., Topeka, KS

GARY HENDERSON, Radian Corporation, Milwaukee, WI

PAUL D. KUHLMEIER, Southern Pacific Transportation Co., Boise, ID

L. JOY LOZANO, Tesoro Environmental Products Co., San Antonio, TX

XUANNGA P. MAHINI, PRC Environmental Management, Inc., San Francisco, CA

MARYELLEN MARTIN, Anson Environmental Ltd., Huntington, NY

STEPHEN G. MCMAHON, Integrated Environmental Solutions Inc., Jacksonville, FL

JOHN W. MELDRUM, Burlington Northern Railroad, Englewood, CO

MARK R. MURPHY, CSX Transportation, Jacksonville, FL

GREG NAUGLE, Hydrologic Consultants Inc., Lakewood, CO

JAMES NOVITSKY, KEMRON Environmental Services Inc., Atlanta, GA

JILL E. RYER-POWDER, ChemRisk Div., McLaren/Hart Environmental Engineering, Burbank, CA

J. TODD STANFORD, Alton Geoscience, Irvine, CA

ROBERT J. STERRETT, Hydrologic Consultants Inc., Lakewood, CO

MICHAEL J. SULLIVAN, ChemRisk Div., McLaren/Hart Environmental Engineering, Burbank, CA

MARLEEN A. TROY, OHM Remediation Services Corp., Trenton, NJ

TIMOTHY P. WIPPOLD, Radian Corporation, Houston, TX

INDEX